LIVING TECHNOLOGY

Other interview books from Automatic Press ♦ $\frac{\text{V}}{\text{I}}$P

Formal Philosophy
edited by Vincent F. Hendricks & John Symons
November 2005

Masses of Formal Philosophy
edited by Vincent F. Hendricks & John Symons
October 2006

Political Questions: 5 Questions for Political Philosophers
edited by Morten Ebbe Juul Nielsen
December 2006

Philosophy of Technology: 5 Questions
edited by Jan-Kyrre Berg Olsen & Evan Selinger
February 2007

Game Theory: 5 Questions
edited by Vincent F. Hendricks & Pelle Guldborg Hansen
April 2007

Philosophy of Mathematics: 5 Questions
edited by Vincent F. Hendricks & Hannes Leitgeb
January 2008

Philosophy of Computing and Information: 5 Questions
edited by Luciano Floridi
Sepetmber 2008

Philosophy of the Social Sciences: 5 Questions
edited by Diego Ríos & Christoph Schmidt-Petri
September 2008

Epistemology: 5 Questions
edited by Vincent F. Hendricks & Duncan Pritchard
September 2008

Probability and Statistics: 5 Questions
edited by Alan Hajék and Vincent F. Hendricks
September 2009

Epistemic Logic: 5 Questions
Vincent F. Hendricks & Olivier Roy
August 2010

See all published and forthcoming books in the 5 Questions series at
www.vince-inc.com/automatic.html

LIVING TECHNOLOGY
5 QUESTIONS

edited by

Mark Bedau

Pelle Guldborg Hansen

Emily Parke

Steen Rasmussen

Automatic Press ♦ $\frac{V}{I}$P

Automatic Press ◆ $\frac{\vee}{\mathsf{I}}$P

Information on this title: www.vince-inc.com/automatic.html

© Automatic Press / VIP 2010

First published 2010

Printed in the United States of America
and the United Kingdom

ISBN-10   87-92130-29-1   paperback
ISBN-13   978-87-92130-29-7   paperback

Typeset in LaTeX2$_\varepsilon$
Graphic design by Milton W. Hendricks & Vincent F. Hendricks

Contents

Preface

Living Technology: 5 Questions

——————————————— ◆ ———————————————

This book is aimed at anyone who is interested in learning more about living technology, whether coming from business, the government, policy centers, academia, or anywhere else. Its purpose is to help people to learn what living technology is, what it might develop into, and how it might impact our lives.

The phrase 'living technology' was coined to refer to technology that is alive as well as technology that is useful because it shares the fundamental properties of living systems. In particular, the invention of this phrase was called for to describe the trend of our technology becoming increasingly life-like or literally alive. Still, the phrase has different interpretations depending on how one views what life is. This book presents nineteen perspectives on living technology. Taken together, the interviews convey the collective wisdom on living technology's power and promise, as well as its pitfalls and perils, from a list of authors with distinguished accomplishments in creating, using, or evaluating living technology.

The editors of this book believe that it is time to start relating and contrasting these lifelike technological systems to the things in the world that we traditionally consider as being alive, because the distinction between the two will increasingly blur in the coming century.

The interviews in this book all address the following five questions about living technology:

- **Question 1: In what sense do you find it meaningful to talk about "living technology?** This question asks whether the term 'living technology' can be given a coherent definition. If so, the next question is whether the term identifies an interesting and important set of things. Different people will use different kinds of standards when answering this question.

- **Question 2: How does your research relate to living technology, and why were you initially drawn to do this work?** This biographical question invites people to explain how they became interested in living technology and started pursuing research in that area.

- **Question 3: How is living technology related to overlapping or nearby research areas, such as nanotechnology, molecular biology, cloning and stem cell research, genetic engineering and synthetic biology? How is it related to social and technological systems such as social networks or information networks, such as the World Wide Web, cell phone networks and electronic banking networks?** This question tries to situate living technology in a broader research landscape, helping us to see what about living technology is similar to earlier technologies and what is new and different.

- **Question 4: What do you think are the most important open research questions about living technology, and how you think they should be pursued?** This question seeks to identify the questions that will most advance research in living technology and may provide a series of interesting starting points especially for graduate students.

- **Question 5: What do you consider to be the most interesting and important human or societal implications of research and development in living technology?** This question solicits people's over-arching visions about the place of living technology in the larger society and what breakthroughs in living technology will mean for society in the coming years.

A rich, multifaceted view of living technology and its prospects can emerge by collecting answers to the same five questions from a wide variety of experts.

When you read through the interviews, you can start to identify certain patterns in what people think about living technology. The contributors to this volume come from a wide variety of backgrounds, including physics, chemistry, philosophy, bioethics, psychology, robotics, engineering, artificial life and artificial intelligence, origin of life research, medicine, sustainable architecture, technology assessment, and research in networking and communications. Most contributors already had a clear idea of what the term 'living technology' means to them, and what it involves in a larger research context. Those that were unfamiliar with the term explore what living technology might mean in the context of their own area of research.

While most contributors already have some sense of what living technology is, they by no means all have the same idea. In order to clearly define living technology, we must have a clear idea of what it is to be alive. Several contributors (see Armstrong, Bedau, Bullock and Peronard) emphasize in their responses to question 1 that we are still lack a consensus on what life is, and what are the fundamental properties of living things. Thus, there is a wide variety the answers to question 1 about the interpretation of "living technology." Some contributors (e.g., Armstrong and Hanczyc) explicitly interpret living technology broadly to include living things used as "technology," i.e., for human purposes (like yeast or cattle), while others (e.g., McCaskill) interpret it more narrowly to include only manmade artefacts that display properties characteristic of biological life. Regardless of how explicit or not they are about precisely which kinds of things should be considered living technology, a recurring theme in answers to question 1 is acknowledgement of the trend, as we move into the 21st century, of technology becoming increasingly complex, powerful, unpredictable and autonomous.

There was also wide variation in type of response to question 3, which asks about how living technology relates to other, similar technologies such as nanotechnology, molecular biology, etc., and how it relates to social and technological network systems. Several contributors (Bedau, McCaskill, Packard and Rasmussen) distinguished in their responses between "primary" and "secondary" living technology: The former is made entirely from nonliving components, while the latter includes some already-living components. This distinction allows primary living technology to have a special status, while secondary living technology can overlap more closely with (or include things from) related areas. Most contributors generally felt that living technology as a research program is highly interdisciplinary, and necessarily involves dialogue with related areas like nanotechnology and molecular biology. Several (e.g., Bedau, McCaskill, and van Est) discuss its close relationship with synthetic biology. Some (e.g., Harvey) felt that it has no relationship with cloning or stem cells, while others (e.g., Packard and Rasmussen) said that these should be included under the conceptual umbrella of living technology. Some contributors distinguished it in various ways from all of the other fields listed in the question. Ikegami, for example, discusses how living technology is different from the others in involving an element of autonomy; Schmidt discusses nanotechnology, molecular biology, etc. as means of implementing the research goals of living technology, once we have cleared the theoretical hurdle of discovering exactly what the relevant living processes are; and Testa discusses living technology as including and building upon work in molecular biology and genetic engineering. Views on the relationship between living technology and social and information networks also varied from the view that living technology will be, at least initially, far removed from the technology of social and information networks (see, e.g., Cronin), to the view that these networks *are* living technology (e.g., Di Paolo), or rather that the increasing autonomous ICT (information and communication technology) components of these networks are becoming more alive (Ulieru), or that they are living technology but only in a secondary sense (e.g., Packard).

Question 4 asks about the most important research questions about living technology. Most contributors indicated that practical research questions relevant to its development and realization are either currently being addressed or should be addressed in the near future, while others emphasized that we have a long way to go before living technology is realizable, and we should focus instead on prior theoretical questions (Schmidt), or even questioned whether we will ultimately be able to produce genuine living technology at all (Støy). A common theme running through many answers to this question was discussion of how creating living technology will help us understand the nature of life (see, e.g., Armstrong, Bedau, and Cronin). Contributors from the robotics field (e.g., di Paolo and Kernbach) often focused on the issue of autonomy – how can we create and control genuinely autonomous systems? Another pattern that emerged in several interviews was emphasis on the question of how living technology will be integrated into the (cultural and physical) environment (see Ikegami, Testa, Peronard, McCaskill, Ulieru, Rasmussen and van Est). Further, Ulieru emphasized the need for human institutions to become more adaptive and agile as part of the development of living technology.

Question 5, which deals with the human and societal implications of research and development in living technology, elicited many responses that focused on the unpredictability of living technology and emphasized that, while we can speculate now about its potential beneficial or the risks it presents, we must wait and see what happens as the technology develops. Opinions on the time scale for confronting these implications varied, with some contributors anticipating breakthroughs in living technology in the next decade or so (Bollen, Kernbach, and Schmidt) and others saying it will likely take longer (see, e.g., Testa). Rasmussen indicated why living technology would likely impact all sectors of our society and how this development could, in the longer term, better integrate and balance human activities with the biosphere and other natural processes, while McCaskill indicated how living technology could resolve some of the societal issues generated by industrialization. Although some (e.g., Peronard) focused only on the benefits promised by living technology, most responses to question 5 addressed the issue of unpredictability and the need to consider both the benefits and the potential risks.

The Initiative for Science, Society, and Policy[1], in Denmark provided the spark that ultimately led to this book, and its content has been especially informed by the ISSP working group on living technology. This group is directed by Mark Bedau, and its members include Rachel Armstrong, Johan Bollen, Rinie van Est, Harold Fellerman, Martin Hanczyc, Pelle Guldborg Hansen, Maya Horst, John McCaskill, Norman Packard, Marco Palombi, Jean-Paul Peronard, Steen Rasmussen, Markus Schmidt, and Kasper Støy. About half of the interviews to this book came from members of the working group. In general, the group is convinced that living technology will be very important in the future of science and society, but the members have different views about what living technology is, what its potential is, how it will evolve in the future, and what its larger implications are for society.

The editors prepared this volume out of the conviction that a collection of opinions from a diverse group of informed individuals would best provide readers with the materials they need to start to make up their own minds about the nature and implications of living technology. Our aspiration is not to provide all the answers, but to help people figure out what questions they might want to ask.

Finally, the editors would like to thank Prof. Vincent F. Hendricks and Rasmus K. Rendsvig without whom the manuscript

---

[1] www.science-society-policy.org

would never have made its final journey into a book.

Mark Bedau, Portland, USA
Pelle Guldborg Hansen, Odense, Denmark
Emily Parke, Philadelphia, USA
Steen Rasmussen, Heraklion, Greece

1

Rachel Armstrong

Co-Director of AVATAR

Architecture and Synthetic Biology, Bartlett School of Architecture, University College London

---

1. In what sense do you find it meaningful to talk about "Living Technology?"

The term 'living technology' embodies a change in the fundamental model of the world and the way that we describe ourselves within it.

There is something qualitatively different about the way that we engage with our technologies in the 21st century compared with the 20th century. This is reflected in how we use them, and is also a property of the technologies themselves, which are pervasive and seamlessly woven into the fabric of our daily lives. Work is no longer performed by people operating systems, as in the original sense of the term 'computer', being a person who was employed to carry out complex calculations; it is carried out instead by digital technology that can work with such speed and accuracy that it does not require human surveillance. Karel Capeck first introduced the term 'robot' in his 1920 play "Rosum's Robots," which described a future where machines replaced people in the workplace. More recently, Ray Kurzweil envisages a future where this technological substitution will be so profound that people will be replaced in terms of their essence, as well as in a functional sense. This "Singularity" culminates in self-aware machines and symbolizes the end of the Anthropocene where people are the dominant species and have actively shaped the nature of the Earth. The advent of technologies that possess some of the properties of living systems but which are not fully alive is a notable development that could suggest that we are traveling along the trajectory described by Kurzweil, and provides a point in technological evolution for reflection on the implications of these apparent differences between 20th- and 21st-century technologies.

The simple observation is that in the 21st century our technologies across the board are more complex. This confers with a set of properties that are hard to describe in a Cartesian way[1] because their modes of operation are interconnected, responsive and emergent. Consequently we experience a new kind of engagement with how we control and interact with these technologies. With pervasive technologies it may not even be possible to see the control point in a system, let alone figure out how to turn it off, shut it down or even remove it. Complexity affects more than just the organizational aspects of these technologies; it is part of their embodiment too. In a Cartesian system there is a hierarchy of order through which it is possible to establish a chain of command. In a complex system however, strategic interventions may not produce the desired outcome and may even generate the converse of what is intended. In practice, the implications are that we no longer have technologies that can be stopped and started with the push of a button. These technologies are able to operate without our direct influence and often even independently of it, their emerging autonomy being one of those "living" features that characterize them.

The use of the term 'living technology' is a necessary step in our engagement with and communication about the science and analysis of complex systems in an organizational and physical manner. We are at a stage where our ability to describe and accurately convey what is happening within a complex system remains a challenge. Visualization techniques are a powerful way of representing complex data, yet interpretation of these graphics is often performed using Cartesian methods of analysis. This is not an incorrect approach, though it does offer an incomplete picture of the complexity of a system under analysis. Until our ways of communicating and articulating complex phenomena within a set of conditions native to complexity are developed it is necessary to find ways of resolving the simplicity of Cartesian description with the richness of complex systems. Living technology can be useful in this context as another viewpoint from which complexity can be described and understood, providing a metaphor that conveys a fundamental shift in our relationship with complex 21st-century technology.

Living systems have frequently been used in the past to solve

---

[1] Cartesian systems are based on objects that are secondarily connected and organized hierarchically, often with a single locus of control.

problems, particularly when machines were not available to perform vital work or where the robustness of animals were more suited to the task. Thinking about how the unique properties of living systems were useful in these contexts helps us understand and articulate what kinds of principles and performance will be relevant to 21st-century complex technologies.

Living technology does not confer a technical measurement of the "aliveness" of a system. An indisputable definition of 'life' does not exist, so it is counterproductive to attempt to use the term as a standardized measuring tool of this complex phenomenon. Yet the term is incredibly persuasive and useful in being able to articulate a set of varying properties that convey something about the nature of the technology under investigation. Living technology can be thought of as a descriptive term, which is relevant to the study of complex systems and can be likened to the Turing Test, a subjective approach to evaluating the phenomenon of intelligence. Alan Turing found it useful to incorporate subjectivity in the assessment of an otherwise difficult to define phenomenon. Similarly, living technology can be best thought of as providing a kind of Turing Test applied to a whole spectrum of technologies by an observer who has an interest in using the properties of living systems as a problem solving tool. The application of the term 'living technology' to this system is then at the discretion of the user whose personal analysis establishes whether the technology passes the "test" in possessing sufficient characteristics to qualify as being useful and indistinguishable from a living system.

In the same way that living systems need to be persuaded, engaged with and maintained in order to get the best out of them, living technologies may also take a while to become established operationally. They evolve over time and may need to be redirected towards a desired outcome. The process is more like gardening than maintaining a car engine. On the one hand this involves a lot of time in nurturing the system, in order to train it to produce the desired outcomes, but the payoff is that the technology itself is more integrated with our lifestyle and our environment. These technologies have the potential to be environmentally responsive and inherently sustainable.

The most uniquely persuasive argument for using a living technology as a problem-solving tool, as opposed to a Cartesian one, is when dealing with uncertainty or unpredictability. These are important properties of real-world systems that have not been "solved" by the application of other technologies. For example, much software has been written for the analysis of complex systems in order to model and rationalize the outcomes of pre-defined sets of variables. In real-world situations, these variables are infinite and there is simply not enough computational power to create fully comprehensive analyses on any given system. This makes traditional computing error-prone when extrapolating models to make an assessment of future scenarios. No matter how sophisticated our programming skills have become it remains impossible to accurately predict the future, and although it is possible to gain a great deal of information about how a system is likely to perform under certain conditions, these future assessments based on software modeling cannot be guaranteed.

Living systems are designed to cope with unpredictability, and are able to maintain contingency with an immediate reality through feedback in forms of sensors and parallel computing techniques. Hewlett Packard is developing a sensory web for computers using real-time sensory input into digital models. This has been termed an Internet of Things, which aims to create a real-time responsive network to environmental events through our computers. Currently the Internet of Things possesses primitive sensors and may, at some stage in the future, be able to respond in real time to changing events, like living systems can. When sensor-modified real-time systems fail to remain relevant in a future context, they no longer exhibit living properties and move towards a state of equilibrium, which we would recognize as "death."

Living systems also have an innate ability to surprise. This is the creative power of living technology, and why it is important that we use it and explore its full potential in the sciences, arts and humanities so that it is possible to meaningfully apply its unique properties to the grand global challenges facing the world today.

2. How does your research relate to living technology, and why were you initially drawn to this work?

Living technology provides an opportunity to imagine and explore how we can use a new set of tools to problem solve differently and examine whether the outcomes are more beneficial than established modes of practice.

My particular interest is in whether it is possible to produce genuinely sustainable architecture using living technology. Urban sustainability is a current grand challenge for the global community with no obvious or easily implementable solutions.

I was drawn to the potential of living technology as potentially giving rise to a portfolio of new approaches that had not previously been possible, and which might bring about significant change in the way that our homes and cities are built. The construction industry is one of the most polluting industries, contributing around forty percent of our carbon footprint. It is based in a practice that uses Victorian tools and technologies, which involves blueprints, industrial manufacturing and construction approaches that employ teams of manual workers. All of this effort requires a one-way transfer of energy from our environment into our homes and cities, to produce objects that are functionally inert. This is anything but a sustainable practice.

Using living technology it may be possible for us to construct genuinely sustainable homes and cities by connecting them to nature, not insulating them from it, so that they can respond to changing needs in populations, alter with the environment and make the most of local resources. These architectures would go beyond the notion of architectural "neutrality" – an idealized notion of architecture using current construction techniques where buildings have no ecological impact on an environment. Using living technology, our cities would become active and participate in processes that could take remedial action on our surroundings, remove pollutants from urban microenvironments and even make useful substances, such as biofuels that could be recycled into the energy-consuming systems of buildings.This would mean that our buildings would be capable of fixing carbon dioxide from the atmosphere and becoming "carbon negative" structures, performing a role similar in some ways to plants.

In order to make this change in our environment, we need the right kind of language to instruct and organize the relevant systems. Living systems are in constant conversation with the natural world, through sets of chemical reactions called metabolism. This is the conversion of one group of substances into another, either through the production or the absorption of energy. My collaborative research aims to develop dynamic, responsive, new materials, termed "metabolic materials," that form the basis for genuinely sustainable architectural tools and methods using a bottom-up approach.

A range of metabolisms is being designed to create a variety of different species of chemical interactions, which are distributed in space and time using self-assembling chemistries as the operating platform. These active systems possess living qualities such as growth, repair and sensitivity to environmental change, which are meaningful in an urban context. Building surfaces would be able to provide an interface for metabolic processes, such as carbon dioxide sequestration and recycling, detoxification and self-repairing materials.

The first metabolic materials are being designed using a dynamic oil-in-water droplet system based on the chemistry of oil and water (Hanczyc et al., 2007). This particular system is able to move around its environment, follow chemical gradients, and undergo complex reactions, but it does not need DNA to instruct its behaviour. It is possible to program it by using general chemical information to influence its outcomes either within the body of the agent itself or by changing the chemical composition of the local environment. Using simple metabolic reactions, the dynamic oil-in-water droplet system can create carbonate, a solid form of carbon dioxide, from solution to produce a shell-like substance similar to limestone, which has been used as a building material for centuries. Using this technology it is possible to produce a material with biological-like properties that can be used in an architectural context. For example, it is possible to create limestone coats or "skins" on the surfaces of buildings so that they become a source of carbon dioxide removal and fixation as well as provide a new kind of building material that can grow on the outside of our homes.

Living technology may provide an alternative approach to some complex problems that have not been effectively resolved in architectural practice. Specifically these are challenges that relate to the sustainability of the built environment and its relationship to climate change.

I was interested in using a chemically programmable dynamic oil-in-water droplet system as a testable model to create architectural outcomes for complex materials. This was particularly interesting since there were no established procedures for control or immediately recognized applications of the technology, and the most important aspect of beginning work with this technology was to find an appropriate problem, in which use of a new system would provide something qualitatively new.

Architecture explores complex ideas and their social and environmental consequences using sophisticated visualizing techniques. This may involve computer modelling, digital fabrication or drawing, and provides an important way of reviewing our understanding of the way that the technology performs and how it may be influenced by its surroundings. Drawings of the oil-in-water droplet system were made based on laboratory observation and speculative propositions about how the system may behave under real-world conditions. Although visualizations are a speculative research tool, the actual process of imagining how a system could perform given a complex set of parameters is a valuable part of the process in understanding what the unique properties of living technologies are and what kind of interventions or "design handles" are necessary to influence the final outcome. Visualizations also serve as a persuasive communications tool through which it is possible to reach a variety of different audiences, from the general public to research grants applications.

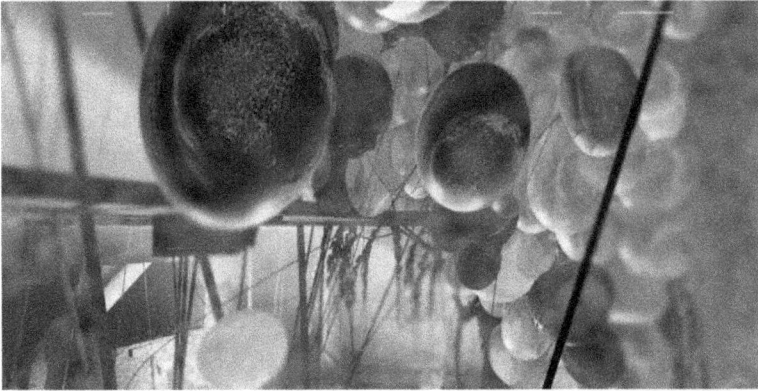

Figure 1. Drawing by Christian Kerrigan of the oil in water droplet technology producing a limestone like substance in a complex real-world environment.

A flagship project using living technology has been designed to examine how these complex materials could influence the long-term sustainability of Venice. Currently, future plans for the city are dependent on the impact of The Moses project, an ambitious engineering project has been proposed consisting of a system of seventy-eight steel floodgates, which will provide a controllable, mechanical barrier at the lagoon's edge to modulate the relentless onslaught of the Adriatic Sea.

From the perspective of living technology, solutions to challenges like this look very different. Thus the construction of an artificial reef under Venice has been proposed using a speculative species of light-sensitive dynamic oil-in-water droplets that are able to move away from sunlight and to produce solid material such as calcium carbonate, which is chemically similar to limestone, when they are settled in their target location. This chemical system would be activated by releasing these programmed oil droplets into the canals, where they would prefer shady areas to sunlight. They would chemically navigate their way towards the darkened areas under the foundations of the city to interact with traditional building materials and turn the foundations of Venice into stone rather than depositing their material in the light-filled canals. These substances can be thought of as being similar to the naturally-occurring substance limestone that has been made from the accretion of skeletons of tiny marine organisms. It is envisaged that this dynamic oil-in-water droplet system will form the basis of a surface coating and repair system for existing building materials that would also be able to remove small amounts of carbon dioxide that have dissolved into the lagoon from the atmosphere. With monitoring of the technology, the woodpiles would gradually become petrified; at the same time, a limestone-like reef would grow under Venice through the accretion and deposition of minerals.

Figure 2. Drawing by GMJ architects of Venice sustainably reclaimed by an artificial limestone reef grown underneath the foundations of the city which spreads the point load of the buildings over the soft delta soil on which the wood pile foundations rest.

The sinking of the city on the soft soil underneath it is just one aspect of the complex set of conditions that beset Venice with its relationship to water levels. The benefits of generating a reef underneath the foundations of the city lie in the way that this reef could distribute the point load of the city on the floor of the lagoon across a hard limestone-like base. Simultaneously, the system could deposit solid material in any gaps forming between the buildings and their foundations, which would further stabilize the city base by extending the solid landmass around the city from the lagoon. Additionally, the reef could reduce the volume of water flowing around the city, thereby buffering it against the effects of water erosion and large movements of subterranean soil as well as attracting the local marine ecology, which would find an ecological niche within this architecture that bridges the built and natural environments.

3) How is living technology related to nearby or overlapping research areas, such as nanotechnology, molecular biology, cloning and stem cell research, genetic engineering and synthetic biology? How is it related to social and technological systems such as social networks or information networks, such as the World Wide Web, cell phone networks and electronic banking networks?

Living technology is not about absolutes. It is liminal, inherently complex and connected and is ideally suited to an interdisciplinary approach. It is also not just about the technology but the way we think and communicate with each other. Living technology becomes a platform through which collaboration is not only possible, but also necessary, to characterize and develop research practices engaged with embodied complexity. Hylozoic Ground, a collaborative project with Philip Beesley for the 2010 architecture Venice Biennale (biannual art exhibition), explores these new approaches and relationships by integrating a responsive, cybernetic framework with complex self-assembling chemistries to produce a public installation that explores a possible NBIC (nanotechnology, biotechnology, information technology and cognitive science) convergence. This collaborative work not only examines living technology on an architectural scale, but also explores the spatial and temporal relationships embodied in the network of integrated systems that comprise the installation, which can be thought of as a "matrix."

The installation consists of a cybernetic framework that may be regarded as a species of living technology, which is embedded in and responsive to perturbations in the environment. This matrix consists of an intricate lattice of small transparent acrylic meshwork links, covered with a network of interactive mechanical fronds, filters, and whiskers.

Tens of thousands of lightweight digitally-fabricated components are fitted with microprocessors and proximity sensors that
react to human presence. This responsive environment functions
like a giant lung that breathes in and out around its occupants. Arrays of touch sensors and shape-memory alloy actuators (a type of
non-motorized kinetic mechanism) create waves of empathic motion, luring visitors into the eerie shimmering depths of a mythi-
cal landscape, a fragile forest of light. Also within the cybernetic
framework are "nested" colonies of dynamic oil-in-water droplet
systems that have been specially engineered for public display so
that they work on a slower and larger scale, and can also respond to the presence of humans through chemistry. This chemistry works on a different time scale, and is complementary to the
digital processors and responsive systems embedded in the cybernetic system. The Hylozoic Ground matrix refers to the collective
array of these connected components, which becomes a site and
substrate for colonization and transformation by a variety of new
materials over a mere fragment of evolutionary time and takes
place during the three-month duration of the site-specific exhibition at the Venice Biennale.

Figure 3. Photograph of the Hylozoic Ground installation by architect Philip Beesley featuring
vials of self-assembling chemical systems that are responsive to local environmental change.

Hylozoism is the presence of life in materials or objects. Hylozoic Ground represents the chemical matter of life as well as living connections created through dynamic physical and chemical processes. The installation is a model system of a synthetic ecology undergoing an evolutionary process through which gallery visitors can observe its initial state through its constituent materials, influence its dynamic processes and review the modifications as an ongoing process for the duration of the exhibition. Notably, the temporal and spatial architectural arrangement of new materials is explored in the installation matrix, which is composed of inert materials that house adaptive chemistries, which may be thought of as a form of living technology, operating through a web of physical, chemical and environmental connections in which the human observers participate. Consequently visitors are able to shape and influence the Hylozoic Ground tissues and organs as well as collectively direct the final configuration of the installation. Within the space and available evolutionary time frame that can be experienced, the change in the composition of the Hylozoic Ground substrate is minimal to the individual visitor, yet hugely significant on a population scale basis, and at a materials level it possesses an apparent inertia. This illusory condition is a function of the multiple scales of complexity at which the system exists, and within the Hylozoic Ground matrix are many subtle, intricate, complex and dynamic ongoing exchanges that confer the installation with living characteristics.

On closer inspection it is possible to observe the constituent processes that connect physical and chemical precursors to the cybernetic framework, which collectively form the matrix. Hylozoic Ground can be compared to a soil structure, which is generally made up of structurally repetitive, organic and inorganic materials that possess heterogeneous properties and conditions analogous with those of tissues and organs in living systems. The various structures are spatially arranged in such a way that they are able to provide suitable surfaces for biochemical exchanges that are compatible with life. This complex matrix of chemicals, materials and environmental fluxes within the Hylozoic Ground oscillate and respond to each other and also to the passage of humans through the model evolutionary installation space, resulting in a series of changes that can be regarded as "synthetic succession." Initially, only a small number of species form *habitats* within the Hylozoic Ground matrix, which are capable of thriving under the initial conditions. The propositional work implies that the adaptive chemistries give rise to spontaneous physical and chemical changes, where new species emerge. These further modify the habitat by altering physical variables, such as the amount of carbon dioxide in the atmosphere, or the mineral composition of the synthetic tissues and organs within the Hylozoic Ground framework. Although within the three-month duration of the installation observers are unable to directly witness a profound evolutionary transformation within the Hylozoic Ground installation, a variety of physical changes are anticipated as a succession of adaptive chemistries. Within the context of the Hylozoic Ground matrix it is possible that given enough evolutionary time, the context for the evolution of new species of living materials, and perhaps new life forms, has been established.

Figure 4. Photograph of adaptive chemistry creating a mineralized structure over time based on the living properties of a dynamic oil in water droplet system

4. What do you think are the more important open research questions about living technology, and how do you think they should be pursued?

The specific research questions that the collaborative installation Hylozoic Ground addresses serve as a model for unanswered questions in the origins of life. Our understanding of this field is enormously incomplete, partly because of our historical position in the scientific understanding of material processes and partly because of the predominately scientific nature of the investigations and theories regarding the origins of life. Whilst Hylozoic Ground exists in a material reality that is most comprehensively described through scientific narratives, its complexity, scale and aesthetics invoke engagement with other disciplines in order to address some key questions more comprehensively. Hylozoic Ground embodies some of the many gaps in our knowledge of the transition from inert to living matter and uniquely attempts to address some pertinent areas that are not fully explained by contemporary scientific debate in order to reach a deeper and more meaningful understanding of living processes, which other disciplines can enrich the knowledge of the field and how life itself arises from terrestrial chemistry.

What is life?
The first challenge posed for gallery visitors by the Hylozoic Ground is for them to make a decision about whether this installation can be thought of as "life," or not. In effect, the participants are being asked to conduct their own Turing Test in a similar manner to the assessment devised by Alan Turing in 1951, which is a comparative analysis of two complex systems. Turing's subject was the phenomenon of intelligence and how it could be successfully measured in the context of "artificial" intelligence. Rather than using traditional decision-making paradigms that require the application of an objective set of hierarchical criteria, Turing exploited observer bias in the decision-making process and asked the observer-participant to decide whether their conversational partner is of human origin, or not. The intelligence source passed the Turing Test if the human participant could not reliably tell the machine from another human. Similarly, Hylozoic Ground asks visitors to make a subjective decision about whether they recognise the adaptive chemistries and cybernetic matrix as being "alive."

What is the architecture of life?
Contemporary scientific discussions regarding the origins of life position the chemistry of modern day cells as being key to understanding the processes that gave rise to the first life forms. Whatever the preferred theory may be that explains the series of molecular assembly processes that were involved at the advent of a biogenesis, the phenomenon is being articulated as two-

dimensional. The traditions of science communication rarefy details of complex analysis within its established conventions, and chemical interactions are therefore conceived of as "cycles" of activity, or as linearly diverging pathways that bear many analogies to the ancient drawings of the Tree of Life. Despite increasingly powerful digital imaging tools, the spatial and temporal (architectural) configurations of adaptive chemistries and the non-linear influences that they exert on these complex processes are simply not possible to articulate comprehensively using contemporary scientific communications tools. Chemistry, even at its molecular level, is so information-rich, capable of many parallel processes and so flexible in its responsiveness to environmental variations, that it is impossible to create accurate models that represent these systems in a naturalistic way. Even the most sophisticated computer models depict information about chemistry that is so rarefied and symbolic in its engagement with its subject matter that, effectively, the analysis remains as incomplete as a two-dimensional linear drawing.

The importance of spatial and temporal information in living systems is already well recognized in scientific experiment. For example, the cytoskeleton (the organizing dynamic framework within a cell) is crucial for the proper functioning of cell processes, and when this is not functioning the vitality of the cell is compromised. This is most dramatically exemplified in reduced gravity environments where micro-gravitational effects on the cell cytoskeleton prevent the development of critical processes in mammalian systems. A meta-analysis in 2006 of experiments carried out by NASA on embryogenesis by Susan Crawford-Young (2006) revealed that the disruption to the mammalian cytoskeletal organization by microgravity has prevented embryogenesis of mammalian life in weightlessness beyond the two cell stage, indicating the profound impact that a disturbance in temporal and spatial organizing structures has for the functioning of complex living systems.

As an interdisciplinary investigation into the temporal and spatial possibilities necessary for the transition from inert to living matter, Hylozoic Ground engages with the "metabolism first" model of organization. This is used to establish a fundamental set of conditions for the various chemical adaptive systems to interact within a framework that supports non-equilibrium states so that the possibility of abiogenesis can be observed. The temporal and spatial positioning of these materials within the matrix is considered to be equally important to the outcome of the installation as the selection of the primary materials themselves. The adaptive chemistries are placed in a landscape where there is thought to be a likelihood that the individual responsive materials will encounter perturbations that have the potential to persist. It is hoped that a fraction of these transformations will give rise to increasingly complex phenomena and ultimately create self-sustaining processes that could be reasonably considered to be "living." The spatiality of the materials is distributed along the kinetic lines of interaction between the Hylozoic Ground cybernetic skeleton where nesting sites for the location of tissue and organ complexes are found. The temporality of the materials is examined as a process of structural and dynamic change of the entire Hylozoic Ground matrix over time and is documented through photography and videography.

Is it possible to model material complexity?
Hylozoic Ground is not a model of a system but embodies the actual set of conditions thought to be necessary for an architecture that could plausibly possess some of the properties of living systems through its chemical, physical, temporal and spatial organization. Hylozoic Ground is complex, highly structured and orchestrated, yet the success or failure of outcomes that are being examined in the space are unknowable from an initial set of conditions, and the most comprehensive analysis of the system is possible only after it runs its course.

Is it possible to model the environment?
Hylozoic Ground does not attempt to model an environment in order to understand the process of organization, but rather opens up the installation space to invite perturbations of chemical, physical or human origin that may facilitate, amplify or agitate change within the participant chemical adaptive systems. Crucially, Hylozoic Ground engages with all aspects of environmental impact as critical processes within the process of abiogenesis and views flux across the matrix as being catalytic in creating necessary, local instabilities within the system. In this way it provides a stark contrast to a traditional scientific approach in running a "control" system through which variability is reduced so that predictable observations can be made. A scientific control subtracts environmental context, peculiarities and incidents from a system, in order to simplify and rarefy information, and establishes governing principles of organization. The Hylozoic Ground provides a matrix in which it is possible to simultaneously observe general principles

of organization as well as examine particular phenomenology in the same context, so that the entire nature of a system that is predisposed to abiogenesis can be appreciated.

How are scale and complexity related?

Hylozoic Ground develops an understanding of how complexity and scale are related through the organizing nuclei that are scattered throughout the matrix as tissue, organs and associated systems, but it also examines transformation of the matrix as a whole within the context of a changing environment. The relationship between scale and complexity of these nuclei is patchy and unpredictable, being localized in time and space. Hylozoic Ground's engagement with the emergence of complexity involves notions of third-order cybernetic teleodynamics,[2] where the interactions of the matrix gain a memory. The Hylozoic Ground structures are influenced by initial conditions and responsive to changes over time, but they also retain information that regulates the behaviors and substrates that can be changed by the memory of the matrix, which give rise to further variations in the performance of the system. In Hylozoic Ground the chemical processes in the installation are under selection pressure from a variety of complex factors that alter the tendency or purpose of the initial system. Despite not possessing any genetics, which is the information processing system which biological systems are classically considered to require in order to experience heritable change, the Hylozoic Ground matrix undergoes various changes that are subject to a dynamic past history. These temporal variations subsequently induce permanent change, and ultimately influence the future behavior and forms of the installation so that the Hylozoic Ground undergoes "teleody-

---

[2] Cybernetics was defined by Norbert Wiener as the study of control and communication in the animal and the machine. The basic theory behind cybernetics proposes that these interactions are controlled by a feedback cycle. When the control engineering and computer science disciplines, which relate to these systems as a sequence of objects, became independent, cyberneticists focused on the internal dynamics of the systems, such as autonomy, self-organization and the role of the observer; this became known as second-order cybernetics. Cybernetics has now moved to the third order, and regards these systems as active-interactive elements participating with each other in complex integrated circuits rather than being a consequence of the action of individual agents. Teleodynamics was formalized in 1984 by the biological anthropologist Terrence Deacon; it is the scientific study of the dynamics that give rise to purposeful behavior, which seeks scientific explanations for the physical origins of intentional behaviors that may be observed in cybernetic systems such as our own.

namic evolution," in other words, evolution that is shaped by the context and processes of the installation that are interacting with the environment and the intentions of the gallery visitors. This process contrasts with the more widely accepted top-down design approach to evolution that employs absolute information and centralized systems, which are characteristic of the information-first theories of evolution primarily implemented by biological design programs, such as genes.

The unexpected

The open system of the Hylozoic Ground invites unpredictable events and engages with them through teleodynamic interactions. Hylozoic Ground is a matrix primed to encounter and engage with unpredictability through the experience of "surprise," since the innate dynamic complexity of the system is unknowable and raises questions about how it is possible to incorporate "the unexpected" into our ways of thinking about materials and processes. Being able to deal with unpredictability is a quality that has not yet been fully articulated or explored in a scientific context and yet this is something that can be experienced by visitors in the context of this installation. Hylozoic Ground embodies fragmentary knowledge reflecting our current understanding of the constituent process of life, but gives viewers access to a much larger terrain that engages with everything that we don't know about the transition from inert to living matter. Hylozoic Ground is partly authored by the constituent adaptive chemistry which has the potential to "fill in the gaps" in our knowledge under the "right" conditions. The Hylozoic Ground matrix raises the possibility of a condition in which the model could be living, through the emergent tangible and irrepressible patterns through which dissipative flows of energy and chemical organization can occur. Since the matrix has the potential to reveal how living processes are organized it is capable of producing surprise in the form of a living entity that can be recognized by visitors, but one that has not previously been encountered by them. Hylozoic Ground can be thought of as a birthing matrix that operates on unpredictable scales of chemical complexity to potentially spit out "life," but not as we have already experienced it.

5. What do you consider to be the most interesting and important human or societal implications of research and develop in living technology?

The implications from the collaborative work relate to the discovery of new technologies, which offer us new problem-solving tools that can alter our role in the world and our influence over it. These new tools may give rise to interventions that enable us to address some of the current grand challenges that have evaded resolution using contemporary approaches.

The city of Venice is the top UNESCO World Heritage Site, and faces destruction from a complex set of environmental and urban conditions. The advent of living technology-based approaches to help reclaim and restore this ancient city with minimal negative environmental impact would signify a new phase in architectural practice, whose lessons and example would be transferrable to many cities that are also facing the consequences of climate change. Equipping Venice with a new set of properties would create a possible future for survival for the city where its very fabric could respond directly to changing conditions, prevent regression into the mud and effectively engage in its own "struggle" for survival, just like living systems do.

Figure 5: Drawing by Christian Kerrigan, Future Venice.

About the Author: Rachel Armstrong is an interdisciplinary researcher with a background in medicine. She is a Senior TED (Technology Entertainment Design) Fellow and Co-Director of AVATAR (Advanced Virtual And Technological Architectural Research), Architecture and Synthetic Biology at the Bartlett School of Architecture, University College London. She collaborates with architects and scientists to create a new experimental space to explore scientific concepts in social settings and re-engage with the fundamental creativity of science. She encourages change and new ways of thinking to develop innovative and different solutions in challenging perceptions, presumptions and established principles around the building blocks of life and society.

References

Crawford-Young, S. (2006).), Effects of microgravity on cell cytoskeleton and embryogenesis. *International Journal of Developmental Biology*, 50, 183-191.

2

Mark A. Bedau

Professor of Philosophy and Humanities

Reed College and

Director

Initiative for Science, Society, and Policy

---

1. In what sense do you find it meaningful to talk about "living technology?"

I think that technology is becoming more lifelike all the time. And I think that the explanation for this is that making technology more lifelike also makes it more powerful and useful. People now imagine technology that is so lifelike that it could be considered to be literally alive. In general, I tend to use the phrase 'living technology' for any technology that is powerful and useful because it is alive or, at least, has certain important properties of living systems.[1] But rather than giving a single definition of living technology, I prefer to catalog its different forms and strengths. To me, there are at least two dimensions along which different conceptions of living technology differ. One dimension concerns whether the technology is *literally alive*, or whether the technology is merely *like life* in certain important respects. I call these *literal* and *analog* living technology, respectively.

The second dimension concerns the kind of life involved in living technology. There are many different views of life in the air today (for a survey, see Bedau and Humphreys, 2008), and they both overlap and differ. For example, some people emphasize metabolism, or evolution, or metabolism and evolution, or metabolism and something else, or other possible combinations of

---

[1] Domesticated animals and hybridized crops count as living technology by this definition, for those organisms are so useful to us because they are alive and because of how we have engineered them over time.

properties. Still others give no details at all about what life is. Accordingly, I distinguish *specific* and *open-ended* living technology, depending on whether the technology involves a specific view of what life is.

This is not an exhaustive account of the kinds of living technology, for there are other dimensions that we could consider. But these two dimensions already define four kinds of living technology. *Literal open-ended* living technology is technology that is alive, literally, but where what it means to be alive is left open. *Literal specific* living technology is technology that is literally alive, where life is defined in some specific way. *Analog open-ended* living technology is technology that is not alive but shares important features of living systems, and it is powerful and useful because of those lifelike features. *Analog specific* living technology is like analog open-ended technology, except that it involves some specific kind or definition of life. Notice that all living technology is powerful and useful precisely *because* it is alive or like life. I don't consider something to be living technology unless the distinctive features of life provide the power behind it.

What someone thinks of living technology is obviously linked to what that person thinks life is. A specific definition of life generates a specific notion of living technology. However, the definition of life is a deep, diverse, complex, and controversial topic. Since life is so controversial, we should expect a diversity of opinions about what living technology is.

I have my own two favorite approaches to defining life, a macro and a micro view. The macro view defines the primary form of life as the process of open-ended evolution, and secondary forms of life, such as individual organisms, are derived from that process in various ways. The micro view concerns the chemical and molecular details of minimal chemical life, and focuses on the Program-Metabolism-Container (PMC) model of life, in which minimal chemical life is viewed as a functionally integrated triad of cooperating chemical systems (Bedau 2010a, 2010c). The macro and micro views of life fit together, and in combination they give a distinctive gloss on what it is that provides the power in living technology. The macro view implies that living technology is powerful most fundamentally because of the open-ended process of creative evolution. The micro view implies that the living technology gets off the ground through the self-assembly and coordinated synergistic control of complex chemical networks and physical structures.

2. How does your research relate to living technology, and why were you initially drawn to do this work?

Looking back, I now see that the central elements of living technology have played a variety of formative roles in the main projects of my professional life. So my answer to this question is lengthy.

I started my professional career as a philosopher, with a PhD from Berkeley on the topic of teleology (explanations involving things like purposes, functions, and goals). I was especially interested in understanding and giving the proper role to the normative or evaluative aspect of teleology.

My view of teleology was implemented early on in the emerging interdisciplinary field called artificial life, or ALife, and led to a family of evolutionary activity statistics and null models that have played a large role in shaping how I think about evolution today. Certain details in the science of artificial life and synthetic biology have subsequently played a central role throughout my professional life. The science has informed many philosophical problems, some novel (strong ALife) and some traditional and revered (the origin and emergence of complex design). As a philosopher, I wanted to contribute to solving those philosophical problems, and I wanted to use the science of artificial life (and, later, other empirical sciences) to inform and shape my solutions, whenever possible. Artificial life computer models have remained a rich source of tools and insights into many of the questions that interested me as a philosopher. The central motivation for spending time on artificial life is perhaps best summed up in Feynman's observation that the best way to show you understand how something works is to be able to build it to your specifications.

A decade ago I became Editor-in-Chief of *Artificial Life*, an MIT Press quarterly journal; this is the central professional journal covering the science of artificial life. I view the field of artificial life as having three overlapping forms, corresponding to three synthesis methods involving software, hardware, and wetware. As Editor of *Artificial Life*, I try to ensure that the journal publishes the full range of the best scientific work in artificial life. My perspective on living technology has been deeply shaped by this work.

At the time I started to edit the *Artificial Life* journal, my colleagues John McCaskill, Norman Packard, Steen Rasmussen, and I embarked on a quest to create a research center that would focus on the fundamental issues concerning living technology. This center was eventually created in Venice, Italy, as the European

Center for Living Technology or ECLT,[2] with the support of the EU-funded FP-6 ITC project "Programmable Artificial Cell Evolution" or PACE, coordinated by John McCaskill. PACE had the overall aim of developing the infrastructure necessary for creating new forms of life in the laboratory. My part in PACE was mostly running a small start-up company with Norman Packard. Our company, called ProtoLife, created the technology for discovering and optimizing complex biochemical systems with desired emergent properties (e.g., properties that are highly synergistic and unpredictable). ProtoLife's technology is one way to engineer systems with these sorts of properties, and achieve what you could call "emergent engineering."

I have a special affinity for the concept of living technology, because John McCaskill, Norman Packard, Steen Rasmussen, and I explicitly formulated that very concept at Ghost Ranch during the summer of 2001 in a proposal for a research center (which eventually morphed and materialized as the ECLT. The four of us explained our shared vision of living technology a year ago (Bedau et al., 2010); this vision has been expanded and sharpened through discussions with many people, including most recently the members of the ISSP working group on living technology. My own views about living technology have been especially profoundly influenced by working with the topic throughout the past decade with John, Norman, and Steen.

When I look back over the topics that have loomed large in my professional life in the past decade, I am struck by how many different threads all involve living technology. Those topics fall under seven overlapping headings, described in the following seven paragraphs. The topics span metaphysics, epistemology, ethics, biology, social science, public policy, and practical commercial activity, but they all centrally concern life or technology in one way or another.

Emergence. I have always been interested in understanding how wholes can be more than the sum of their parts. This kind of emergence is a characteristic hallmark of all forms of life. Today there is a rebirth of work by philosophers and scientists on emergence (see Bedau and Humphreys, 2008). My own contribution to this work has emphasized the "weak" emergence found in mechanisms that are so complex that their behavior is unpredictable (Bedau, 1997, 2010d). The only way to learn how weakly emergent systems

---

[2] See http://www.ecltech.org.

will behave is to watch them and wait. Weak emergence is found throughout the study of complex, highly synergistic systems, like synthetic cells, and it will be a characteristic hallmark of all living technology.

Evolution. I am fascinated by the power and creativity of the evolutionary process that shapes all actual life forms. Natural selection played a pivotal role in my dissertation on teleology, and also in two decades of experiments with computer models of evolving populations of agents. The evolution in these models was weakly emergent, as far as anyone knows, so to learn their generic properties we performed empirical parameter sweeps. In this way I studied a family of evolutionary activity statistics, devised and modified over the years with Norman Packard and a team of other colleagues and students. These statistics have been very useful for visualizing and quantifying adaptive evolutionary dynamics in a variety of evolving systems. One main upshot of this work has been the realization that no artificial life model today demonstrates interesting forms of open-ended evolution.

The evolution of technology. Our inability to model interesting forms of open-ended evolution made me wonder what makes biological evolution so open-ended. I think that one mechanism behind open-ended evolution is *door-opening innovations*, which occur when the evolution of one form of life opens the door to new opportunities for other life forms. The same mechanism seems to shape the evolution of technology, for a key invention can open the door to many families of new technologies. For example, the invention of the ink-jet printer opened the door to printing metal compounds, printing intricate three-dimensional plastic structures, printing DNA on glass slides to make micro-array chips, and even printing skin cells to make tissue for skin grafts. One can apply activity statistics to the evolution of technology, by using the patent record. The patent record is a beautifully accurate and complete record of the creation of almost all new technologies. Measuring the impact on innovation shows that the superstar drivers of innovation in the past thirty years are PCR, ink-jet printing, and stents, among others.[3] It also shows that many of the biggest technology drivers are door-opening inven-

---

[3]Each of these innovations was revolutionary. PCR is one of the cornerstones of contemporary biotechnology. I explained above how ink-jet printing enabled the precise positioning in three-dimensional space of microscopic bits of many kinds of matter. Stents enabled coronary heart surgery without opening the chest cavity.

tions (Buchanan et al., 2010). One can visualize the evolution of the technology drivers as a cloud of high-content words from patent titles and abstracts, sized by use (Chalmers et al., 2010). From this perspective, the whole technosphere[4] is evolving in a way that is analogous to the kind of evolution that happens in the biosphere. In particular, I suggest that door-opening innovations make a significant impact on new inventions in both biological and technological evolution. If the technosphere is evolving in an open-ended way, driven by door-opening innovations, then it is analogous to an evolving biosphere. This means that the technosphere itself is a kind of living technology. The components of the technosphere include many nonliving technologies (individual cell phones, etc.) and also many living human beings, who are intentionally creating and using the individual cell phones, and the like.

Protocells and synthetic biology. The PACE project introduced me to the details of the science of "protocells" – that is, the attempt to make living systems using only minimal kinds of nonliving chemical ingredients. Protocells and the other constructions from recent synthetic biology are especially important contemporary examples of living technology. Protocells and synthetic cells are some of the simplest known chemical forms of life, and research in this area experimentally explores the transition zone between life and nonlife. I helped create a roadmap for protocell research (Rasmussen et al., 2009), which centrally figured the Program-Metabolism-Container (PMC) model of minimal chemical life and presented Rasmussen diagrams to depict the functional chemical structure of different kinds of protocells, both those achieved in the laboratory and those merely proposed. My recent philosophical work on the nature of life makes central use of both the PMC model and Rasmussen diagrams.

The nature of life. My philosophical training combined with my involvement in artificial life naturally led me to think about the nature of life. I have come to the conclusion that life is an expression of some of the most fundamental processes in nature. Those processes are very powerful and useful, and they involve many interesting properties like creativity, novelty technology, innova-

---

[4]The technosphere is the network of all exant technologies in their actual environments and the people who use them. In the technosphere people are small, individual players caught up in a larger web, one that is changing and adapting continually.

tion, adaptation, and evolution, among others. I have argued for the central role of open-ended evolutionary creativity in the nature of life (Bedau, 1996, 1998), and I have used the PMC model of minimal chemical life to explain in what sense minimal chemical life is a matter of degree (Bedau, 2010a, 2010c). The view that the nature of life raises and the way it informs many philosophical and scientific questions led me to produce two recent collections about life (Bedau and Cleland, 2010; Bedau 2010b).

Engineering emergence. The technology that Norman Packard and I created with the team at ProtoLife allows you to automate the discovery and optimization of biochemical systems with desired emergent properties. This technology uses statistical models built from experimental data to optimize the design iterations of high-throughput experiments. A few dozen iterations of this cycle automate an intelligent and highly efficient empirical exploration of the emergent properties distributed in huge and complex experimental spaces. Discovering and optimizing the emergent properties of complex biochemical systems is a holy grail today in synthetic biology, in biotechnology in general, and in living technology in general.

Social and ethical implications of biotechnology. I am concerned about the broader social and ethical implications of my scientific work, especially the work in synthetic biology that involves creating new forms of life in the laboratory. Similar concerns arise for many forms of living technology. The PACE project produced a series of workshops at the ECLT on broader perspectives on protocell research, a book of new essays on the moral and social implications of creating life in the laboratory (Bedau and Parke, 2009), and a table of social and ethical checkpoints for protocell research and development (Bedau et al., 2009). Following the inspiration of Steen Rasmussen and the leadership of the University of Southern Denmark, I have recently helped create the Initiative for Science, Society, and Policy (ISSP), in Denmark.[5] One of ISSP's current projects concerns living technology; that project has produced this book of answers to five questions about living technology.

3. How is living technology related to overlapping or nearby research areas, such as nanotechnology, molecular biology, cloning and stem cell research, genetic engineer-

---

[5] See www.science-society-policy.org.

ing and synthetic biology? How is it related to social and technological systems such as social networks or information networks, such as the World Wide Web, cell phone networks and electronic banking networks?

To answer this question, it is useful to remember the distinction between primary and secondary living technology (Bedau et al., 2010).[6] *Primary* living technology is made from components that are all nonliving, while *secondary* living technology is made from some components that are living. The synthetic cell created by Venter's team (Gibson et al., 2010) and other constructions of top-down synthetic biology are secondary living technology, by this definition, because they start with a natural living organism (usually, a bacterium) and modify it by giving it a prosthetic genome. Another example of secondary living technology, by this definition, might be the social network that exists on Facebook, provided this network exhibits enough of life's defining features. If this social network is itself alive it is secondary living technology, because the network includes a large number of living people.

These two examples of secondary living technology are very different. Venter's synthetic cell is the synthetic modification of an *internal part* of a *single* instance of a natural individual organism of the *simplest* natural life form (a bacterium). The modification merely reproduces the *old powers* of the natural organism out of which the synthetic cell is made.[7] By contrast, the social network is a complex *external network* of technology that connects *many* instances of the *most complicated* natural life form (people). The resulting technological network gives its members many *new powers* that do not exist without the network. There are many different ways to create hybrid technologies that critically involve living organisms, and secondary living technology is a grab-bag

---

[6] I believe that we need to tweak the definition of primary/secondary living technology. Web social networks should be primary living technology, even though some of their components (human beings) are alive, and even though the whole Web works only as long they remain alive. The point is that the fundamental lifelike properties of the whole system (the Web) do not merely borrow the lifelike properties of the components. By contrast, that is exactly what Venter's synthetic cell does. It critically depends on the complex biological black box that is the host bacterium, and this bacterium is a normal living organism. So, Venter's artificial cell is secondary in a way that social networks on the Web are not.

[7] Future synthetic cells could presumably have many new, useful features added, but only if we have somehow solved the problem of engineering emergence (see following text).

that includes various kinds of hybrid systems that involve natural living organisms as critical components.

In addition to thinking about individual technologies that are like life to some degree, we should also think about the network of all the technology in the world and all of the people who use it. I call this network the technosphere, on analogy with the biosphere. The evolution of the technosphere and the biosphere are alike in some ways and unlike in others. For example, they each involve populations of entities that interact, compete and cooperate, and the composition of the population changes over time. A process like natural selection seems to happen in certain niches in the technosphere, when variant technologies compete for market share. However, the source of variation in the population is quite different in the two processes. Biological variation comes from random mutation and crossover, and a mixture of other processes such as random genetic drift and genetic linkage. Technological variation is the product of the intentional, conscious, top-down, intelligent choice of human designers. These and other differences aside, for me the interesting question is whether the evolution of technology is like biological evolution in some fundamental abstract way, that is deeply connected with its power and creativity.

4. What do you think are the most important open research questions about living technology, and how do you think they should be pursued?

The study of living technology understood broadly encompasses at least the study of synthetic biology, autonomous robots, and distributed autonomous communication networks. Each of these areas of research, and each of their subfields, has its own important open research questions. So, rather than attempt to be synoptic, I focus only on key open research questions that loom largest in my own work: the seven research areas described in my answer to question 2 (above). These questions are quite varied: Some are scientific, some involve engineering, some have direct commercial impact, and others are metaphysical, epistemological, ethical, and political.

In protocell science, the overriding scientific milestone is to produce the first fully synthetic cell. The synthetic cell created by Venter's team (recall above) is only about 1% synthetic, measured by dry weight. Creating a synthetic cell that is synthesized 100% from nonliving chemical materials is the holy grail in protocell science. Since minimal chemical life in the PMC model is defined functionally (Rasmussen et al., 2009; Bedau, 2010a), a fully synthetic cell in principle could be chemically quite different from all familiar biological life.

Though many will readily agree that the first fully synthetic cell is a huge scientific milestone, the protocell research community is still in need of a clearer and more comprehensive list of important scientific milestones. Formulating and publishing such milestones is an important constructive task to accomplish. A list of analogous milestones for creating lifelike social-technical systems is another important scientific milestone itself.

In metaphysics and epistemology, a central open question is: What is the right theory of the evolution of technology? There are a number of different points of view on this topic, including the constructivist toolbox view of Brian Arthur (Arthur, 2009) and the natural selection of cultural items depicted by memetics (Dawkins, 1976/1989; Dennett, 1996; Aunger, 2001). It is important to distinguish two questions about the connection between living technology and the evolution of technology. One question is whether people have created or will create technological devices that are alive, when considered individually. These individual technologies that are alive are the exception; virtually all technology today is not alive. The second question concerns not individual technologies but the technosphere as a whole, which links people with the technologies that they make or use. This is the question whether the technosphere is or could be alive, even if none of its component technologies are alive when considered individually.

A related fundamental scientific and philosophical question is to discern the right way to understand the nature of life. I think we should answer this question with details about the most important and powerful properties of living systems, and the underlying mechanisms that produce them.

These open philosophical questions are connected to an important open practical question: How can we engineer complex, highly synergistic systems to have the unpredictable emergent properties that we desire? I call this the problem of *engineering emergence*. ProtoLife's technology is one approach to engineering emergence. It is difficult to engineer life's useful properties because they are emergent properties. The molecular-level causal processes that generate them are very complicated. Many nonlinear lower-level factors interact synergistically. This makes the desired properties emergent and unpredictable, and thus expensive to discover and optimize experimentally.

ProtoLife's solution to this problem is to use state-of-the-art statistical models and prediction algorithms to automate an intelligent and highly efficient experimental search in a huge and complex space of possible experiments. The technology has already produced results. One involved engineering a family of new liposomal drug formulations (Caschera, Gazzola et al., 2010); drug formulations are complex chemical systems with many components (e.g., elements of a lipid library) that interact nonlinearly and synergistically in ways that cannot be predicted from first principles. Another involved improving the efficiency of one of Invirogen's cell-free protein synthesis kits (Caschera, Bedau et al., 2010).

5. What do you consider to be the most interesting and important human or societal implications of research and development in living technology?

I think that we will make technology with more and more of life's fundamental properties, because those properties will enable our technology to be much more powerful and useful. This prospect raises new forms of old questions about the positive and negative effects of technology on people. Will living technology be pernicious and dangerous, and something that we should avoid, or will it be more personalized, natural, intuitive, and less alienating than older, traditional nonliving forms of technology? Both are possible, I think, and which direction we go depends partly on whether and how those working on living technology step up to their scientific social responsibility. Some of these responsibilities are merely specific instances of responsibilities borne by anyone who creates powerful new technologies and commercializes them. Other responsibilities are very closely tied to certain specific important properties of living technology, such as having emergent properties.

Life's emergent properties have a profound impact on the social and ethical implications of living technology for at least two reasons. First, if a given piece of technology were alive, then its own behavior would be unpredictable, although experience in the lab and the field could quickly empirically teach us a lot about what to expect. Second, even if individual pieces of technology are not themselves alive, perhaps the whole technosphere is alive. That is, in addition to any issues raised by individual technologies becoming alive, there is an additional issue concerning the lifelike behavior of the whole technosphere.

When deciding the future of living technology, we are deciding in the dark. This means that traditional risk analysis does not work. We need to rethink how to make public policy when deciding in the dark. Caution is especially important when deciding in the dark, which is one reason for the contemporary appeal of the precautionary principle (surveyed in Parke and Bedau, 2009). But even when deciding in the dark, caution should be tempered with other virtues such as courage (see Bedau and Triant, 2009).

As with all profoundly new technologies, there is no way to accurately predict the full implications of living technology; we must simply wait and watch the future unfolds. That is one consequence of the emergent properties throughout life. Systems with emergent properties are often too opaque for us to foresee all the dangers and risks they will create in the future. We should acknowledge this, and take appropriate steps (such as learning how to engineer complex systems with desired emergent properties—ProtoLife's solution is outlined in my answer to question 4). I think that we never foresee all the important consequences of any important new technology, including such things as automobiles, PCR, ink-jet printing, stents, and the mobile phone, but also stem cells and synthetic biology. If the different forms of living technology that people actually create are so hard to predict, then even given all that we know about the relevant science, how can we possibly find a responsible course of action for today? One answer is to tabulate which specific actions are triggered by which scientific milestones. As the science progresses, we can fill out more details in the table and focus our attention on different actions.

These efforts seem to be normal expressions of scientific social responsibility. A self-conscious effort to enact and promote scientific social responsibility is the core mission of the Initiative for Science, Society, and Policy (ISSP), which was created in Denmark through the vision and energy of Steen Rasmussen and many others. ISSP alerts and informs all interested parties (scientists, stakeholders, policy makers, businesses, ethical experts, and the general public) about the social and ethical issues concerning science and technology. ISSP also takes steps to enable the relevant science have a constructive influence on our society and its policies. Through these kinds of positive influences, the ISSP has brought about the context that has led to this book.

About the Author: Mark A. Bedau (Ph.D. Philosophy, UC Berkeley, 1985) is Professor of Philosophy and Humanities at Reed College and a regular Visiting Professor at the European School of Molecular Medicine in Milan, Italy. He is an internationally recognized leader in the philosophical and scientific study of living systems and has published and lectured extensively on issues concerning emergence, evolution, life, mind, and the social and ethical implications of creating life from nonliving materials. He has published over 100 research papers and co-authored or co-edited 7 books, including *Emergence: Contemporary readings in philosophy and science* (MIT Press), *The prospect of protocells: Social and ethical implications of creating life from scratch* (MIT Press), and *The nature of life: Classical and contemporary perspectives from philosophy and science* (Cambridge University Press). He is Editor-in-Chief of the journal *Artificial Life* (published by MIT Press), co-founder of ProtoLife Inc., co-founder of the European Center for Living Technology (Venice, Italy), and co-founder and director of the Initiative for Science, Society, and Policy (www.science-society-policy.org).

References

Arthur, B. (2009). *The nature of technology: What it is and how it evolves.* New York: Free Press.

Aunger, R. (Ed.) (2001). *Darwinizing Culture.* New York: Oxford University Press

Bedau, M. A. (1996). The nature of life. In M. Boden (Ed.), *The philosophy of artificial life* (pp. 332-357). New York: Oxford University Press.

Bedau, M. A. (1997). Weak emergence. *Noûs,* 31 (supplement 11), 375-399.

Bedau, M. A. (1998). Four puzzles about life. *Artificial Life,* 4, 125-140.

Bedau, M. A. & Humphreys, P. (Eds.) (2008). *Emergence: Contemporary readings in philosophy and science.* Cambridge: MIT Press.

Bedau, M. A. & Parke, E. C. (Eds.) (2009). *The ethics of protocells: Moral and social implications of creating life in the laboratory.* Cambridge: MIT Press.

Bedau, M. A., Parke, E. C., Tangen, U., & Hantsche-Tangen, B. (2009). Social and ethical checkpoints for bottom-up synthetic biology, or protocells. *Systems and Synthetic Biology,* 3, 65-75.

Bedau, M. A. & Triant, M. (2009). Social and ethical implications of artificial cells. In M. A. Bedau & E. C. Parke (Eds.), *The ethics of protocells: Moral and social implications of creating life in the laboratory* (pp. 31-48). Cambridge: MIT Press.

Bedau, M. A. (2010a). A functional account of degrees of minimal chemical life. *Synthese*, forthcoming.

Bedau, M. A. (Ed.) (2010b). *Synthese*, special issue on philosophical problems about life.

Bedau, M. A. (2010c). An Aristotelian account of minimal chemical life. *Astrobiology*, forthcoming.

Bedau, M. A. (2010d). Weak emergence and context-sensitive reduction. In A. Corrandi & T. O'Conner (Eds.), *Emergence in science and philosophy* (pp- 46-63). Routledge.

Bedau, M. A. & Cleland, C. E. (Eds.) (2010). *The nature of life: Classical and contemporary perspectives from philosophy and science.* Cambridge: Cambridge University Press. Forthcoming.

Bedau, M. A., McCaskill, J. S., Packard, N. H. & Rasmussen, S. (2010). Living technology: exploiting life's principles in technology. *Artificial Life,* 16, 89-97.

Buchanan, A., Packard, N. H., & Bedau, M. A. (2010). Adaptive innovative impact on the evolution of technology in the patent record. In S. Rasmussen et al. (Eds.), *Proceedings of Artificial Life XII.* MIT Press, forthcoming.

Caschera, F., Bedau, M. A.,Buchanan, A., Cawse, J., de Lucrezia, D., Gazzola, G., Hanczyc, M. M., & Packard, N. H. (2010). Coping with complexity: Machine learning optimization of cell-free protein synthesis. Preprint.

Caschera, F., Gazzola, G., Bedau, M. A.,Bosch Moreno, C., Buchanan, A., Cawse, J., Packard, N., & Hanczyc, M. M. (2010). Automated discovery of novel drug formulations using predictive iterated high throughput experimentation. *PLoS ONE,* 5(1), e8546. doi:10.1371/journal.pone.0008546

Chalmers, D., Cooper Francis, C., Pepper, N., & Bedau, M. A. (2010). High-content words in patent records reflect key innovations in the evolution of technology. In S. Rasmussen et al. (Eds.), *Proceedings of Artificial Life XII.* MIT Press, forthcoming.

Dawkins, R. (1976/1989). *The selfish gene.* Oxford University Press.

Dennett, D. C. (1996). *Darwin's dangerous idea*. New York: Simon and Schuster.

Gibson, D. G., Glass, J. I., Lartigue, C., Noskov, V.N., Chuang, R.-Y., Algire, M. A. Benders, G. A., Montague, M. G., Ma, L., Moodie, M. M., Merryman, C., Vashee, S., Krishnakmar, R., Assad-Garcia, N., Andrews-Pfannkoch, C., Denisova, E. A., Young, L., Qi, Z.-Q., Segall-Shapiro, T. H., Calvey, C.H., Parmar, P.P., Hutchinson III, C.A., Smith, H. O., & Venter, J. C. (2010). Creation of a bacterial cell controlled by a chemically synthetized genome. *Science*, 329, 52-56.

Parke, E. C. & Bedau, M. A. (2009). The precautionary principle and its critics. In M. A. Bedau & E. C. Parke (Eds.), *The ethics of protocells: Moral and social implications of creating life in the laboratory* (pp. 69-87). Cambridge: MIT Press.

Rasmussen, S., Bedau, M. A., McCaskill, J. S., & Packard, N. H. (2009). A roadmap to protocells. In S. Rasmussen, M. A. Bedau, L. Chen, D. Deamer, D. C. Krakauer, N. H. Packard, P. F. Stadler, eds., *Protocells: Bridging nonliving and living matter* (pp. 71-100). Cambridge: MIT Press.

3

Johan Bollen

Associate Professor

School of Informatics and Computing, Indiana University

---

1. In what sense do you find it meaningful to talk about "living technology?"

It is impossible to discuss the notion of "living technology" without making reference to existing thinking on what constitutes biological life. When we attempt to explicate what we mean to say when we deem an entity alive we are faced with the important distinction between making extensional and intensional definitions. In the former (extensional) we define a concept or category by the exemplars that are considered representative of that concept or category. For example to define the concept of "bird" I could provide a set of known birds. In the latter type of definition (intensional) we define a concept or category by outlining the features and properties that entities must have to be considered valid members of the concept or category. For example, to define what we mean by "bird" we can specify that it is a species whose members must have wings and feathers and lay eggs.

In a similar vein we can define the concept of life, or the category of living entities, either by means of a set of exemplars that we consider to be representative of life in the aggregate, or by formally stating the properties that entities must conform to for them to be considered alive. The former approach is fruitless given the incredible variety of biological life that spans from the most rudimentary virii and protocells to self-aware, creative organisms such as the higher mammals and homo sapiens. The properties of living things (in the intensional sense) are therefore usually defined in functional terms, i.e., as the ability to reproduce, evolve, adapt, metabolize, grow, and self-regulate. Such definitions, although not universally accepted, manage to cover much of what we would intuitively consider biological life (with a few interesting exceptions). However intensional definitions of life have an interesting side effect. Their focus on the functional, general characteristics of life greatly broadens the set of entities that could conceivably be considered alive. A variegated multitude of hypothetical and implausible organisms could conform to these functional characteristics yet be far removed from what we would intuitively consider alive.

The attempt to define "living technology" follows directly from this issue. If it makes sense to discuss life in terms of its basic properties and categories, equally it makes sense to examine cases in which existing or hypothetical non-biological forms are endowed with the same general properties.

Since the industrial revolution technology has generally been considered distinct from biological life. Whereas some technology could adapt and self-regulate, none of it could conceivably reproduce, evolve or metabolize. In addition, the information-processing capabilities of most technology lacked the highly adapted, sophisticated nervous systems that were commonly found in higher life forms and supported life's ability to adapt and self-regulate.

Recent advances in materials science, nanotechnology, information technology, computing, and theoretical biology, however, are all pointing towards an inflection point in technological progress. Bedau et al. (2010) explains the imminence of living technology and make an interesting distinction between primary and secondary living technology, i.e., technology that is endowed with the properties and capabilities of biological life, and either consists entirely of nonliving components or aggregates living, biological components into functional units that as a collective can be considered alive. Resorting to extensional definitions of living technology one could conjure the potential of artificial protocells, self-assembling and reproducing chemical reactions, self-healing materials, and online ecologies.

Of particular interest in this discussion are the now ubiquitous social networking environments in which chat bots, users, social networking software, games, productivity tools, recommender systems, and virtual reality environments are aggregated into complex ecologies that may at some point in time be considered living technology. This form of "virtual" living technology is less subject to physical constraints and thus more prone to rapidly and freely evolve into systems of considerable complexity that can be studied in their own right, and could in fact elucidate many of the still open research questions surrounding biological life and living technology in general.

In my view the notion of living technology has moved from speculation to an imminent next step in technological and societal evolution. As a subject with tremendous societal, scientific, and theoretical implications it merits an interdisciplinary research and development program that moves beyond technological and ethical speculation.

2. How does your research relate to living technology, and why were you initially drawn to do this work?

My lifelong interests and academic training have focused on how online networks and adaptive technology shape human behavior, social behavior and communication. I graduated with a MS degree in Psychology from the University of Brussels (VUB) in 1994, specializing in autonomous robotics. I started my PhD studies that same year under the direction of Francis Heylighen[1] whose interests in cybernetics, systems science and societal evolution matched my own in social networks and adaptive information systems. These were the early days of the Web, and Francis along with Cliff Joslyn (PNNL) and the late Valentin Turchin (City College and CUNY) had founded the Principia Cybernetica Project (PCP), an online initiative to create an evolving ontology of systems science.[2]

The structure of the PCP ontology was intended to evolve under the direction of its board of editors, but we soon began to speculate on evolving, autonomous knowledge networks. Our interests in dynamic information systems and collective intelligence led to the formulation of the first conceptualizations of the Global Brain idea; the notion that the Earth's communication networks, and in particular the Internet, form a central nervous system for an evolving social and economic super-system that emerges from the interactions between individuals and technology.[3] The Global Brain, as defined by Francis and I in 1994-1996, represents the prototypical notion of a living technology, and is the foundation of my interests in this domain (Heylighen, 1996; Bollen, 1996). During my tenures at Los Alamos National Laboratory, Old Dominion University and Indiana University I have continued to explore the role of technology in creating and sustaining systems that leverage

---

[1] See http://pespmc1.vub.ac.be/heyl.html.

[2] See http://pespmc1.vub.ac.be/.

[3] See http://pespmc1.vub.ac.be/suporgli.html.

the power of collective intelligence (Heylighen, 1999; Rocha, 2000; Bollen, 2002; Rodriguez, 2007; Bollen, 2007). My most recent interests have focused in particular on studying the flow of ideas in scientific communication and the spread of sentiment and opinion through online social networks. Both areas are increasingly influenced by living technology thinking.

My interests in living technology are quantitative and empirical. I do not wish studies of living technology to be limited to philosophical, ethical and moral explorations of future technology. It is crucial that empirical science and engineering converge on tangible problems in this developing domain so that we expand both the science of living technology as well as its development and practice.

3. How is living technology related to overlapping or nearby research areas, such as nanotechnology, molecular biology, cloning and stem cell research, genetic engineering and synthetic biology? How is it related to social and technological systems such as social networks or information networks, such as the World Wide Web, cell phone networks and electronic banking networks?

Clearly the study of living technology will be a highly interdisciplinary endeavour that will interact with a variety of scientific domains concerned with biological life and complex networks and systems. The question pertains to what distinguishes the study of living technology from existing efforts. I believe the answer lies in one important distinction.

Living technology is specifically focused on studying instances of technology that are endowed with the properties and characteristics generally associated with biological life from a systemic point of view, i.e., across a wide variety of possible instantiations. As such it is not necessarily concerned with the engineering or design of artificial life or intelligent materials, but rather with the generalization of concepts from the study of life in general to technology. This generalization enables the study of lifelike technology across a wide variety of domains that can stretch from the creation of synthetic protocells to the study of online environments whose principles of organization and functionality approximate those of biological life.

In this sense I again believe a great opportunity exists in the realm of social and technological systems where physical analogies to biological life are less relevant than the overarching systemics, and a functional perspective can be employed to study general problems of reproduction (in a zero-cost reproduction environment), homeostasis, and autopoiesis.

4. What do you think are the most important open research questions about living technology, and how you think they should be pursued?

The notion of living technology revolves around what is essentially an empirical claim; it formulates a so-called falsifiable hypothesis, namely, that some technology can be so fully endowed with the characteristics commonly attributed to life that it is in fact a form of life. I believe this claim needs to be examined, not merely by argument, community consensus or from theory, but as an empirical and engineering problem. This will require a research program not dissimilar to that concerned with the detection of extraterrestrial life; life that has not yet been observed but whose properties we induce from what we know about terrestrial life, and which we attempt to detect on the basis of a variety of chemical, behaviorial and functional markers.

Our community needs to establish a set of operationalizations of markers of living technology that will allow the formulation of specific null hypotheses that form the underpinnings of empirical work in this area. For example when we observe the properties and behavior of online social networking systems or in fact engineer a specific online environment, we should not merely be content with drawing analogies to biological life. We should carefully define a set of operationalizations of specific markers of living technology, define subsequent null hypotheses with regards to such networks being alive, and proceed to empirically falsify or confirm such hypotheses. This type of work will bring forth concrete progress in this domain on all levels, including its philosophical and theoretical underpinning as well as the engineering of living technology. In fact, the empirical validation of the claim that particular technology can be considered living will have profound implications akin to the detection of extra-terrestrial life.

I therefore believe the most important open research questions is how we define the empirical markers of living technology, and how to detect living technology and observe its properties and characteristics. These three research questions constitute a conditio sine qua non for this domain.

5. What do you consider to be the most interesting and

important human or societal implications of research and development in living technology?

Living technology will have a profound effect on our society as a whole and on us as individuals, possibly to the same degree that the introduction of agriculture changed human culture and society. The human and societal implications of living technology will, in my opinion, be most pronounced for digital, online systems that can conceivably reproduce, evolve, change and self-regulate without the restrictions that their physical counterparts would be subjected to. They will therefore be at the vanguard of advances in living technology, and will cause living technology to have the most pronounced effect on how we communicate and socialize online.

Within the next 5 or 10 years I do not necessarily see a revolution, but a consistent trend towards online systems of increasing intelligence, self-regulation and adaptability. These systems will allow individuals to maximize their ability to find relevant information, analyze and synthesize this information, and manage their personal social environments. Eudemonic feedback loops may result that allow individuals to tune their goals, aspirations and activities to achieve greater personal and social efficiency. The components of such living technology are presently emerging in online services that provide individuals feedback on fluctuations of their mood (as detected from their online activities), their diets, their social networks, and provide assistance in organizing one's personal information infrastructure. The productivity and well-being of individuals may be profoundly impacted by the convergence of these life management systems.

In societal terms I envision the emergence of similar living technology that allows societies as a whole to better self-regulate economic activity, improve the efficiency of social programs, and better manage their political and governmental activities. These systems may very well develop along the lines described in the Phenomenon of Science by the late Valentin Turching (Turchin, 1977), and the Principia Cybernetica Project. The notion of a Metasystem Transition (MST) pertains to the evolution of metasystems that establish new levels of control by integrating and controlling the interactions of the underlying systems, increasing the overall system's adaptability, intelligence and efficiency. According to this theory, metasystem transitions have come about through evolution several times in the history of life, in particular in the emergence of multicellular life and of cognition as an additional level of control in multicellular organisms. Each metasystem transition adds an additional level of control that brings about new levels of intelligence and complexity. MST theory speculates that the next metasystem transitions could occur at the societal level. Our economic and communication infrastructure could evolve a societal metasystem that organizes the interactions of individuals, technology, communication infrastructure and even our governmental systems. It could as such be considered the prototypical case of living technology. Whether such metasystem transition has yet occurred and whether it is desirable is a matter for debate, but my thinking of the social impact of living technology is very much aligned with this school of thought. I encourage those interested in the study of living technology to familiarize themselves with this body of work.

About the Author: Johan Bollen is associate professor at the Indiana University School of Informatics and Computing. He was formerly a staff scientist at the Los Alamos National Laboratory from 2005-2009, and an Assistant Professor at the Department of Computer Science of Old Dominion University from 2002 to 2005. He obtained his PhD in Experimental Psychology from the University of Brussels in 2001 on the subject of cognitive models of human hypertext navigation. He has taught courses on Data Mining, Information Retrieval and Digital Libraries. His research has been funded by the Andrew W. Mellon Foundation, National Science Foundation (NSF), Library of Congress, National Aeronautics and Space Administration (NASA) and the Los Alamos National Laboratory. His present research interests are usage data mining, computational sociometrics, informetrics, and digital libraries. He has published extensively on these subjects as well as on matters relating to adaptive information systems. He is presently the Principal Investigator of the Andrew W. Mellon Foundation- and NSF-funded MESUR project, which aims to expand the quantitative tools available for the assessment of scholarly impact.

References

Bedau, M. A., McCaskill, J. S., Packard, N. H., & Rasmussen, S. (2010). Living technology: Exploiting life's principles in technology. *Artificial Life, 16,* 89-97.

Bollen, J. & Heylighen, F. (1996). Algorithms for the self-organization of distributed, multi-user networks. In R. Trappl (Ed.) *Proceedings of the 13th European Meeting on Cybernetics and Systems Research* (pp. 911–917), Vienna, Austria. Austrian Society for Cybernetic Studies.

Heylighten, F. & Bollen, J. (1996). The world-wide web as a super-brain: From metaphor to model. In R. Trappl (Ed.), *Cybernetics and Systems '96* (pp. 917–925). Austrian Society for Cybernetic Studies.

Heylighen, F., Bollen, J., & Riegler, A. (Eds.) (1999). *The evolution of complexity.* Dordrecht: Kluwer Academic Publishers.

Rocha, L. M. & Bollen, J. (2000). Biologically motivated distributed designs for adaptive knowledge management. In I. Cohen & L. Segel (Eds.), *Design principles for the immune system and other distributed autonomous systems* (pp. 305–334). Oxford: Oxford University Press.

Bollen, J. & Nelson, M. L. (2002). Adaptive networks of smart objects. In J. Bollen & M. L. Nelson (Eds.), *Proceedings of the Workshop on Distributed Computing Architectures for Digital Libraries (ICPP2002)* (pp. 487–496), Vancouver, B.C., Canada, August 18-21 2002. IEEE.

Rodriguez, M. A., Steinbock, D. J., Watkins, J. H., Gershenson, C., Bollen, J., Grey, V., & deGraf, B. (2007) Smartocracy: Social networks for collective decision making. In *Proceedings of the International Conference on Systems Science (HICSS)*, Hawaii, January 2007.

Rasmussen, S., Mangalagiu, D., Ziock, H., Bollen, J. & Keating, G. (2007). Collective intelligence for decision support in very large stakeholder networks: The future US energy system. *ALIFE 2007-IEEE Symposium on Artificial Life*, April 2007.

Turchin, V. (1977). *The phenomenon of science: A cybernetic approach to human evolution.* New York: Columbia University Press.

4

Seth Bullock

Head of Science and Engineering of Natural Systems Group

University of Southampton

---

1. In what sense do you find it meaningful to talk about "living technology?"

It strikes me that this question has two related interpretations: what is the significance of the research activity organising around the term 'living technology', and what issues are raised by the notion of living technology as a concept or category. First, I think that there are many different valid responses to these questions, and that mine are largely coloured by the fact that my work is far removed from the actual attempt to synthesise living technologies. For me, the significance of that enterprise lies in the challenge that it presents to the design, engineering, control and management communities – a radically new perspective on what technology can be, on how technologies can be built, and on how we are expected to interact with them. It seems to me that the perspective on systems engineering presented by living technology research represents a very timely new paradigm that is sorely needed.

As for deciding what things should and should not count as exemplars or instances of living technology: Since we are dealing with a notion that has yet to develop into a consensually defined mature category, we will not always know whether particular systems should be deemed "in" or "out" (for two examples, see question 4, below). However, this is nothing to be ashamed of or worried about. I started out in cognitive psychology and artificial intelligence, and got used to working on a set of problems that are also hard to pin down: perception, memory, reasoning, intelligence, mind, and consciousness. Consensually agreed mature definitions that operationalise these terms rather than simply delineating a set of phenomena to be explained just don't exist yet. Achieving them is the end of the story, not something to be established before the story can begin.

So my short answer to the opening question would be "in the widest sense possible." I tend to subscribe to a generous reading of "living technology" because I am interested in the boundary cases, whether they are man-made or naturally occurring, organic or inorganic, or even socio-technological. My guess is that most contributors to this volume are involved in the attempt to *bring about living technology*, which might understandably mean that they are more interested than me in tightening up the definition ahead of time. Possibly they feel that some synthetic biotic machines either already fall into the category of living technology or will soon. Most may feel that simple, tiny, lifelike systems built from non-biological materials will eventually join them. After all, to feel otherwise is tantamount to the assumption that "life's core properties" will forever be possessed only by systems made from already living materials. The same practitioners may concede that, technically, there are also naturally occurring, *non-synthetic* examples of living technology, where whole animals or plants are employed as machines either by us (e.g., bees, yeast) or even each other (e.g., parasitic flukes that directly influence the brain of their ant host in order to control its movement); but it is likely that these are not their primary concern. More controversial are socio-technological systems that may include living entities as parts, but are also claimed to exhibit "life's core properties" at the level of the whole system, e.g., hospitals, governments, universities, online communities, etc. For some, my guess is that this may be a step too far, but personally I would welcome them in too.

I don't feel the need to strongly defend the claim that any of these kinds of systems are *really* examples of living technology, or that they *really* are not, since my interest in the field stems precisely from the way in which it will shed light on the nature of the "core properties of living systems" that lie at the heart of both the living technology enterprise and my own research interests.

2. How does your research relate to living technology, and why were you initially drawn to do this work?

Well, I see two core reasons for working on living technology: first, to synthesize useful new technologies that solve important problems for society, and second, to better understand what the core properties of living systems are by generating new examples of technologies that purport to exhibit these properties. The former is an increasingly significant activity, but my primary interest is in the latter, and I see the living technology field as an important new way of stimulating this project.

As I mentioned, my background is in psychology and artificial intelligence (AI). Like others involved in what became known as "nouveau AI," my personal trajectory has been ever downwards. Within cognitive science, I climbed down from the heady heights of my undergraduate degree studying human logic, language, rationality and reason, and writing AI algorithms designed in the image of clinical, Kasparov-like ratiocination, in order to pursue a PhD in Sussex's Evolutionary Robotics lab. There I was confronted by the crude, clumsy effectiveness of robots inspired by tiny insects, driven by toy neurons, automatically wired together by witless evolutionary algorithms: AI from the bottom up. In parallel, I was working with experimental economists. There, I dutifully worked through the prescriptive Bayesian analyses of rational choice theory, but then plumbed the depths of real people's inability to string two rational decisions together in a risky choice experiment. As my PhD took a biological turn, I coded up Dawkins' eerily simple algorithm for abstracting the raw power of natural selection freed from the messy particularities of terrestrial biochemistry, and then struggled to understand the manifest gap between the resultant insipid simulated evolution and the full-blown complexity of the real thing.

In each case, I was left feeling that the realities of life, mind and society were far removed from the clean algorithms, theorems and calculations of classical AI, neo-classical economics and neo-classical Darwinian biology. Their abstracted, formal accounts were attractively clean, elegant and economical, but this "neat" approach was fundamentally misleading. What was missing seemed obvious to me at the time. I was embedded (and embodied and situated) in a research environment suffused with a Heideggerian conviction that all behaviour happens *in the world*, that it is "embedded, embodied, situated," that Simon's emphasis on "satisficing" rather than "optimising" was the appropriate attitude, and that Kauffman's "order for free" could be the bedrock for adaptive behaviour.

There is not space here to do much justice to the ideas that I've just name-checked, but I will indicate roughly what brings them together for me, which is a thoroughgoing naturalisation of life and mind. Martin Heidegger's philosophy challenges a still pervasive Cartesian mindset that takes thinking and reasoning to be essentially general, rational, logical, computational and (tacitly) magical, in that they are assumed to take place in an idealised mental realm of symbols, logico-syntactic structure-sensitive functions, etc. For Heidegger, cognition, thought, language, logic and ultimately *being* must all be recognised as situated in the world no matter how angelic and transcendental they might appear. In taking this turn he opens the door for an entirely new enactivist cognitive science that prioritises a concern with the coupling between creatures and their environments as much as their sensing and acting or beliefs and desires, and does this for creatures irrespective of whether they can be imagined to possess an internal "language of thought." (See Wheeler, 1995, 2005, for more on the link between Heidegger, cognition and artificial life.)

Thirty years after the publication of Heidegger's (1927) *Being and Time* (but a decade before its first translation into English), Herb Simon (1957) articulated his idea of *bounded rationality*. He operationalised a notion of grounded cognition, pointing out that it is not in the interest of an organism to reach optimal solutions to the problems posed by its environment if the time or resources consumed in doing so are prohibitive. Decision making is always *situated* in an evolving environment that is specific to the decision maker. The dynamics of this *Umwelt* are critical to understanding an organism's behaviour, cognitive or otherwise. (See Goodie et al., 1999, for more on the threads of bounded rationality that run through cognitive science, biology and economics.)

If the Heidegerrian attitude of Sussex's evolutionary robotics group was appropriate for cognitive behaviour, it was also surely true of life more generally. Just as the locus of intelligent behaviour was not an algorithm running inside a head, the locus of "vitality" was not to be found in a free-floating evolutionary algorithm comprising heritable variation plus differential reproduction. Stuart Kauffman's (1993) *Origins of Order* set out a series of results that demonstrated how useful, complex biological organisation might arise in the absence of natural selection, and that this ordered organisation was more than just the backdrop or raw materials for evolution. Rather, facts about terrestrial physics and chemistry had a substantive part to play in how adaptive processes had played out on Earth and would continue to do so.

The challenge, then, was to understand what aspects of this dirty, frustrating *embeddedness* were involved critically in underpinning the phenomena of life, mind and, ultimately, society. Modelling was the route that I took to exploring these questions, building simulations in which to explore the role of "logistics" (who does what to whom, where and when) in effective adaptive biological organisation. How does the spatial embedding of a population influence natural selection (Clark and Bullock, 2007)? How does a population's environment structure influence the costs of irrationality (Bullock and Todd, 1999)? How do logistics influence the evolvability of signalling (Noble et al., 2001)?

Along with the construction of real-world living technologies, simulation models of this kind are an example of the *synthetic methodology*. They are attempts to understand systems not by taking them apart and exploring the properties of the pieces in the reductionist hope that systems' secrets are located in the properties of their atoms, but by synthesising systems bottom-up. By assembling a system's parts together in a computer or a petri dish or on a workbench, the aim is to explore the relationship between the organisation of these parts and the properties of the whole that they form.

It seems to me incontrovertible that brute facts about the nature of the implementation level will be key to understanding the behaviour of complex adaptive systems and that we cannot therefore continue to abstract these facts away in neat, tidy theorem-friendly formalisations. Self-organisation, thermodynamics, spatial embedding, etc. are key to unlocking the secrets of how brains, cells, organisms, and communities work when they do. Attempts to synthesise and understand living technologies directly and necessarily confront these implementation issues. My hope is that in doing so they will shed new light on the problem of what underpins living behaviour in much the same way that examining patients with brain damage sheds light on regular cognitive behaviour. Note that even in cases where such brain damage results in abnormal cognition, or in behaviour which is not "cognitive" at all, they are valuable to the study of cognition, since exploring where and how a system breaks down is a good way of finding out how it was working in the first place. Analogously, whether or not we end up deciding that a particular new technology is "truly living" is to a large extent irrelevant to how much it might teach us about what it is about the world that brings systems to life.

3. How is living technology related to overlapping or nearby research areas, such as nanotechnology, molecular biology, cloning and stem cell research, genetic engineering and synthetic biology? How is it related to social and technological systems such as social networks or informa-

**tion networks, such as the World Wide Web, cell phone
networks and electronic banking networks?**

From my perspective, the most interesting boundaries are between
the synthesis and study of living technologies and our understanding of what makes certain biological, social and socio-technological
organisations *vital*. What makes one hospital *click*, while another
consumes massive amounts of energy and people in delivering a
mediocre service? What makes one community resilient to insults
and shocks, while another is fragile or moribund? How could buildings or transport networks become systems that organise themselves creatively in a symbiotic relationship with their users?

In some sense, artificial synthetic cells are just a set of exciting new analogies for existing complex adaptive systems – just
a new set of vocabulary with which to gloss the same ideas and
questions. In my experience it is easy to undervalue such new
ways of speaking, since they lead to new ways of thinking. But in
this case there is a second and potentially more significant cross-
disciplinary contribution to be made by living technologists: the
struggles, failures and successes in living technology labs will provide real experimental data on the nature of autopoiesis.

Unlike my simulation models, where the abundant degrees of
freedom offered by a modern computer ensure that almost anything can be made to hold within a digital world, synthesis in the
real world is massively constrained by physics and chemistry. It is
clear that what works must work despite these constraints. From
my perspective, the fascinating possibility is that living technologies will work not despite these constraints, but *because* of them –
i.e., the constraints of physics and chemistry will be *enabling* for
life (Bullock and Buckley, 2009).

**4. What do you think are the most important open re-
search questions about living technology, and how you
think they should be pursued?**

There are obviously many complicated problems that remain to
be solved in realising examples of living technology in the lab, and
other contributors to this volume are far better qualified to assess
them than myself. So I will limit myself to raising one issue that
is perhaps further down the line but will need to be addressed before autonomous, adaptive, free-living technologies are employed
in earnest. How will these technologies adapt during their use?
How will we ensure that this adaptation is benign? How could

we in fact exploit the adaptive power of living technologies rather than seek to attenuate it?

Here, I would suggest that there may be value in studying naturally occurring living technologies – living technologies that were not deliberately engineered by people, but arose spontaneously in nature and have persisted and adapted ever since.

Two (candidate) examples of non-anthropogenic living technology (that have perhaps not been recognised as such) have occupied me in some of my work: termite mounds (Ladley and Bullock, 2005) and biological signalling systems (Bullock and Cliff, 1997).

Termite mounds are clear examples of technology in that they are carefully constructed homes, they are just not our homes, not homes that are planned or engineered, but instead self-organise. If we consider the mound and the termite colony that built it together as a single system, then there are grounds for considering it as an example of *living* technology (in much the same way that the Internet is sometimes taken as such). The structure itself is sophisticated and multi-functional, serving as shelter and protection from weather and predators, but also as a functionally segregated environment with specialised areas in which to care for offspring, bury the dead, raise crops, etc. The mound is adaptive and homeostatic, maintaining critical parameters such as chamber temperature and humidity via self-regulatory air flows, and integrity via self-repair and ultimately self-reproduction. However, to grant this system the status of living technology is complicated by the role of the living beings that produce it – aren't they the only living aspect of the system, with the mound simply being a product of their activity? While this is a legitimate perspective, it may be that we will not fully understand the nature of termite mounds, how they adapt and their stigmergic relationship with their inhabitants, until we take a perspective that recognises them as autopoietic living systems in their own right.

Across the natural world signalling systems are rife: from the messenger molecules employed by the simplest cells to the syntactically structured utterances of the most complex primates. Signalling occurs for a multitude of different reasons and takes a huge variety of different forms: from songs and calls to smells, gestures, postures, and patterns. All this sharing of information takes place despite the competition to survive and reproduce. How do natural signalling systems arise, persist, and adapt? Why do signals take the form that they do? To some extent these questions can be answered by treating signalling systems as living technology. Signals are tools for transmitting information and their form reflects this function. The temporal structure of some bird song, for example, is adapted in a way that allows it to resist degradation by reverberation in forested habitats and thereby travel further. Simon Kirby's group has shown that properties of human language may arise as a consequence of different language variants competing to successfully be transmitted through the bottleneck of language learning in a human infant (Kirby, 2002). While signalling systems are not *engineered* technology, and they are not even *substantial* in the sense of a piece of physical hardware, nevertheless it may be the case that a full understanding of them requires us to consider them as evolving adaptive living technologies.

In both cases, we can view these techno-social systems as comprising an organismal population that produces, supports and maintains a technological superstructure. However, it is also possible to see causal processes that run in the reverse direction. Technology clearly impacts directly on the organisms that use it, sheltering them or informing them, but the selective processes that these organisms are subject to also slowly shape them, fitting them to an environment in which the technology is a dominant feature. In some sense then, a natural language comes to direct the way that its speakers think and see the world, and a termite mound *builds and maintains itself* by steering the short-sighted behaviours of its termite slaves.

If we are to enter a design space populated by living technologies of the kind described above, or other kinds, we will need to understand how these co-adaptive relationships unfold, and be able to steer them when necessary.

5. What do you consider to be the most interesting and important human or societal implications of research and development in living technology?

In a world that cannot continue to sustain us in our current resource-hungry mode of existence, we must quickly learn how to build and use systems that *self-organise* to deliver the quality of life that we need. Living technology is one route to such a future. If civilisation's job is, very broadly, to lower entropy in our local environment at the expense of increased entropy out in space somewhere, then technologies driven by self-organisation, if designed correctly, will offer us solutions that require minimum energy in order to deliver the ordered world that we need. By contrast, our traditional approach to technology is analogous to Maxwell's demon, carefully creating order by deliberately, manually shifting stuff from one place to another, but at the expense of consuming the massive amounts of energy and time required to measure and control the world, and resist and recover from the natural processes that are going on around us. Living systems demonstrate that we can do better. A few genes steer a process of self-organisation to create a creature, perhaps a person. A sub-set of these genes steer processes of self-organisation to create a learning brain. A further sub-set steer processes of self-organisation in order to create a living cell operating at efficiencies unheard of in human engineering, in environments that are massively challenging, and at collaborative scales that are currently unthinkable.

About the Author: Dr. Seth Bullock gained both his first degree and PhD at the University of Sussex in what was then their School of Cognitive and Computing Sciences (COGS), first studying cognitive science and then researching evolutionary simulation modelling. After spells working in Berlin and Leeds, where he founded the Biosystems research group, he now heads the Science and Engineering of Natural Systems (SENSe) research group within the School of Electronics and Computer Science at the University of Southampton, and is founding Director of the University's new Institute for Complex Systems Simulation. He has a history of interdisciplinary research into a wide range of different complex systems, and has core interests in modelling methodology and evolutionary theory. He was Conference Chair for the Eleventh International Conference on Artificial Life on its first visit to Europe in 2008, and sits on the board of the International Society for Artificial Life. He publishes in journals spanning health-care, economics, biology, computing, architecture and physics, and was the only engineer invited to contribute to Richard Dawkins' 2006 OUP festschrift.

References

Bullock, S., & Buckley, C. L. (2009). Embracing the tyranny of distance: Space as an enabling constraint. *Technoetic Arts*, 7(2), 141-152.

Bullock, S., & Cliff, D. (1997). The role of 'hidden preferences' in the artificial co-evolution of symmetrical signals. *Proceedings of the Royal Society of London, Series B*, 264, 505-511.

Bullock, S., & Todd, P. M. (1999). Made to measure: Ecological rationality in structured environments. *Minds and Machines*, 9(4), 497-541.

Clark, B., & Bullock, S. (2007). Shedding light on plant competition: Modelling the influence of plant morphology on light capture (and vice versa). *Journal of Theoretical Biology*, 244(2), 208-217.

Goodie, A. S., Ortmann, A., Davis, J., Bullock, S. & Werner, G. M. (1999). Demons versus heuristics in artificial intelligence, behavioral ecology, and economics. In G. Gigerenzer & P. M. Todd (Eds.), *Simple heuristics that make us smart* (pp. 327-355). Oxford: Oxford University Press.

Heidegger, M. (1927/1962). *Being and time*. Trans. J. Macquarrie & E. Robinson. London: SCM Press.

Kauffman, S. (1993). *Origins of order: Self-organization and selection in evolution*. Oxford: Oxford University Press.

Kirby, S. (2002). Natural language from artificial life. *Artificial Life*, 8(2), 185-215.

Ladley, D., & Bullock, S. (2005). The role of logistic constraints on termite construction of chambers and tunnels. *Journal of Theoretical Biology*, 234, 551-564.

Noble, J., Di Paolo, E. A., & Bullock, S. (2001). Adaptive factors in the evolution of signalling systems. In A. Cangelosi & D. Parisi (Eds.), *Simulating the evolution of language* (pp. 53-78). Heidelberg: Springer.

Simon, H. (1957). A behavioral model of rational choice. In *Models of man, social and rational: Mathematical essays on rational human behavior in a social setting*. New York: Wiley.

Wheeler, M. (1995). Escaping from the Cartesian mind-set: Heidegger and artificial life. In F. Moran, A. Moreno, J. J. Merelo, & P. Chacon (Eds.), *Advances in artificial life: Proceedings of the third European conference on artificial life* (pp. 65-76). Heidelberg: Springer.

Wheeler, M. (2005). *Reconstructing the Cognitive World: The Next Step*. Cambridge, MA: MIT Press.

5

Leroy Cronin

Gardiner Professor of Chemistry

Department of Chemistry, University of Glasgow

---

1. In what sense do you find it meaningful to talk about "living technology?"

I feel that living technology, from a materials point of view, outside of biology or computing science is still very much an unproven concept, and therefore I often think and discuss the ideas or consequences of living technology very much in eager anticipation of what could be possible.

For me the key aspect of any living technology is the potential for autonomous adaptation. Coupling this property with present day engineering paradigms opens up a vast world of material processes and new devices; from a scientific point of view this is incredibly interesting, since it combines the approaches of design and evolution together with the idea of autonomous and "adaptive matter." This implies that the entire process of evolution of the matter is done in the chemosphere (chemical world). Of course, human designers already practice the process of adaptation (1st- to 4th-generation Apple iPods, etc), but they do so as "observers," tweaking parameters using a reductionist type approach that cannot possibly sample all possible environmental aspects simultaneously in proportion to their effect on the design.

Living technology is still a very abstract concept with only one material example, that of biology. I think the concepts and ideas embodied by living technology go way beyond biology, but this potential has not yet been made into a "hardware" reality (beyond simulation and weak adaptation via software simulation and embodiment in robotics). Of course, living technology approaches exist (or are simulated) *in silico*, and indeed are harnessed for real-world applications (e.g., machine learning and genetic algorithms). In the "real" world, the manipulation and engineering of biology is resulting in the emergence of the field of "synthetic biology" (which is really a rehash of systems biology), but I do not feel it appropriate to talk about "living technology" in the context of synthetic biology. This leaves some of us with a problem, since if we only count biology, and do not count synthetic biology, where does that leave discussion on living technology, outside of a computational or simulation context?

I think the answer is simple, and will guide much of the research and development needed to realize this in the "materials" world; this answer is the idea or paradigm that I would like to call "evolvable matter." I refer to evolution here in a strict Darwinian sense, not in the abstract increasing sophistication (i.e., increasing fitness against a variety of design specifications) that is implied by an observer with no knowledge of the information content or origin of the information that defines the material – e.g., defining evolution as the survival of the fittest with a process by which adaptation and information transfer from one generation to the next can result in a fitter system. This strictness is important, since I think the idea of how we discuss "living technology" defines a very important question. That is, is evolution outside of biology a universal property or "law," or is biology the only route by which matter can evolve? I think it is interesting to phrase the question in this way since it allows one to explore a radically different perspective.

Although I may have expressed these ideas in a rather abstract sense, I think the following examples will serve to illustrate the point practically. If evolution can be engineered to occur with "raw" unsophisticated "matter," then we well may be able to evolve sophisticated materials with properties simply not accessible via normal chemistry routes. For example: (1) using the idea of superconductivity as a fitness function in our evolvable material platform could allow us to improve our superconducting materials so that room temperature superconductivity is possible. (Superconductivity is the process by which electrical energy can move through a conductor with zero loss due to heat, but right now we can only do this at rather chilly liquid nitrogen temperatures.) (2) Drug evolution by defining a receptor-small molecule interaction as a fitness function. Improving this using evolutionary synthetic techniques in the organic laboratory could allow the "evolution" of perfect drugs. The introduction of unwanted interactions causing side effects could be also set as obstacles for the process to navigate around.

Solving these problems using an evolutionary approach in the pure chemical world (i.e., without biology) may allow a new, slightly less controversial, definition or working hypothesis of what life or living systems embody. In our own work we were struck by the many philosophical and scientific issues associated with an all-encompassing definition; we have suggested that a Turing-like test may help researchers work towards systems that we could all agree show some common attributes associated with that of living systems, i.e. work in the laboratory towards more sophisticated life like systems (Cronin et al., 2006). Perhaps I could even suggest that it is impossible to robustly discuss what "living technology" is in the absence of such an approach, since the physical, philosophical and even metaphorical arguments would lack meaning or relative calibration.

Even if the "Turing test" for life fails (e.g., if an observer could not tell the difference between artificial and natural living systems, or perhaps even more intriguingly, an artificial creature could be accepted as being alive by a global "living" community), it is still important to grapple with global characteristics of living systems (see Figure 1). For example most, if not all, living systems are characterized by their own morphological persistence over many generations. I.e., an immense amount of work is done by the living "system" to maintain form, function, and evolution of that form through a number of cycles. Living systems require energy and entropy evolution, and somehow characterize the arrow of time in a way that fundamental physics is not able to do (although maybe cosmology, e.g., star birth, death, etc. does in this respect). This also could be expressed as "living systems tend towards kinetic stability" whereas "chemical systems tend towards thermodynamic stability" (Pross, 2005). However, the concepts of birth, evolution and death are all in the world of the biochemical. Quantum theory and classical mechanics are invariant with respect to the direction of time (that is, the equations equally make sense with time reversal), whereas living processes require entropy to increase (for the universe; i.e., they are dissipative). Therefore, discussing living technology also may yet yield a profound new approach to understanding thermodynamics, energy, and information. Indeed, perhaps it is not really possible to discuss living technology with any certainty without a real definition of what physical minimal attributes life requires.

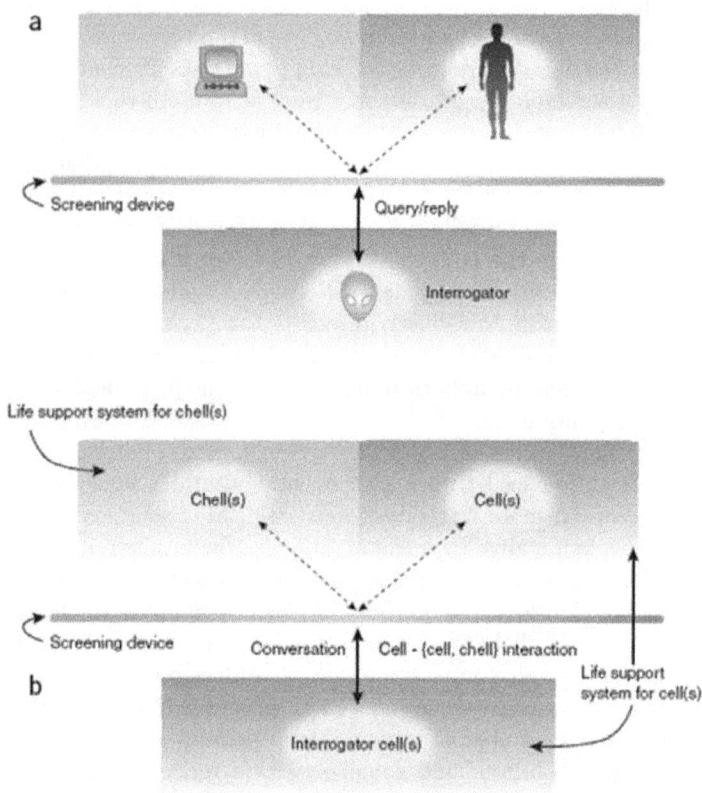

Figure 1. Different takes on Turing. (a) Representation of the classic Turing test with an intelligent interrogator (that is, a person, interacting with two compartments, each containing either a computer or a person, the location of which is unknown to the interrogator). (b) The extension of this interrogator or interaction between cells and chells (see text) is depicted. It is suggested that cell signaling could be used as the medium for conveying the interactions, but other, perhaps simpler, mechanisms also could be used (see Cronin et al., 2006).

2. How does your research relate to living technology, and why were you initially drawn to do this work?

I am doing research in adaptive chemical systems, including the creation of systems that allow evolution in materials using a machine-assisted approach, and this project (funded by the UK Engineering and Physical Sciences Research Council (EPSRC), entitled "Evolvable Process Design") seeks to use a "top-down" enforced evolutionary approach upon a materials system to make more sophisticated matter (c.f. evolution of a room temperature superconductor as described in Question 1). In this respect, we are trying to bootstrap the ability to simulate biological-like evolution computationally into the inorganic world. Beyond this, in another EPSRC-funded collaborative project entitled "The Chell," we are trying to work towards chemical cells or "Chells" that can help us understand if a chemical cell is the fundamental building block of most living systems (of course, slime moulds go beyond this) (Pasparakis et al., 2010). Further, I am interested in seeing if inorganic chemical-cells or "iChells" can be used to simulate and re-create an inorganic origin of life (Cairns-Smith, 2008), and if we can create a type of inorganic, artificial biology (rather than synthetic biology) based upon a different set of elemental building blocks. I chose this project to see if the "rules" are generic (i.e., whether the rules of chemical complexity leading to living systems are the same as found in biology); one positive aspect of this research is that if our creation escapes, it will not be viable in the current environment (so it is much more self-limiting than GM or synthetic biology). Also, chemically speaking, informally I am yet to come across a fundamental chemical limitation that could limit the assembly of life in anything more than a few thousand hours from the elemental soup that was present on Earth 500 million years after its birth. The process of evolution after the living chemical infrastructure is established is another thing altogether. Curious to see if there is a fundamental limitation in what we are attempting, we recently established a new project in my laboratory (http://www.croninlab.com) to see if we can "create life in 5000 hours," the typical lifetime of some bright Xeon lamps.

I am drawn to this work since we are asking some very fundamental questions using a chemical toolbox. What is the origin of life on Earth? Is life on Earth unique? Can we create evolvable matter outside of biology? What is the minimum chemical infrastructure required for replicating (Kindermann et al., 2005) "living matter" to be sustained? I.e., is the separation of roles in biology (e.g. DNA, proteins, sugars, cell membranes, metabolic pathways, ATP, etc.) a generic role distribution required for all "living" systems?

The requirement of the minimum chemical infrastructure to establish a complex system is absolutely key, and a missing link in much research that touches on the origin and creation of life. Although our current understanding of biology has allowed the creation of the fields of biotechnology and genomics in a very short time, there is really no firm evidence that yields a clear view with regard to the origins of biology. One of our key research project aims to change that by at least demonstrating 2-3 routes by which life can emerge on Earth from the inorganic world (as described above). I am sure about one thing: The origin of life was inorganic (this is a rather obvious conclusion given the definition of organic "living" systems). For instance, in my Science paper published in January 2010, we showed that the assembly of protein-sized inorganic metal oxide clusters templated by a smaller oxide cluster could be done in a flow system whereby the overriding chemical controls were acidity and the presence of a reducing agent (see Figure 2) (Miras et al., 2009).

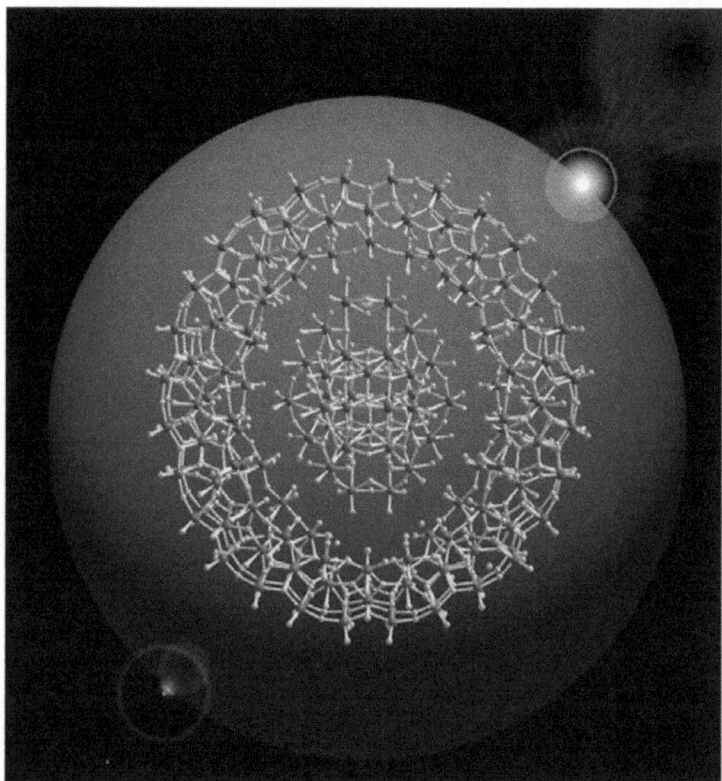

Figure 2. Depiction of the structure of the gigantic molybdenum oxide "nanowheel," complete with a "transient" guest that was trapped inside the wheel under chemical-flow-conditions (see Miras et al., 2009).

Also, in other fundamental work on the self assembly of large nanoscale metal oxide molecules, we have demonstrated that the structure, topology and nuclearity of the clusters can be controlled quite precisely by chemical parameters as simple as acidity (pH) (Miras et al., 2008). Indeed, in some of these systems, pH can be seen as the information-carrying part of the system that defines one molecular architecture, as opposed to another (see Figure 3).

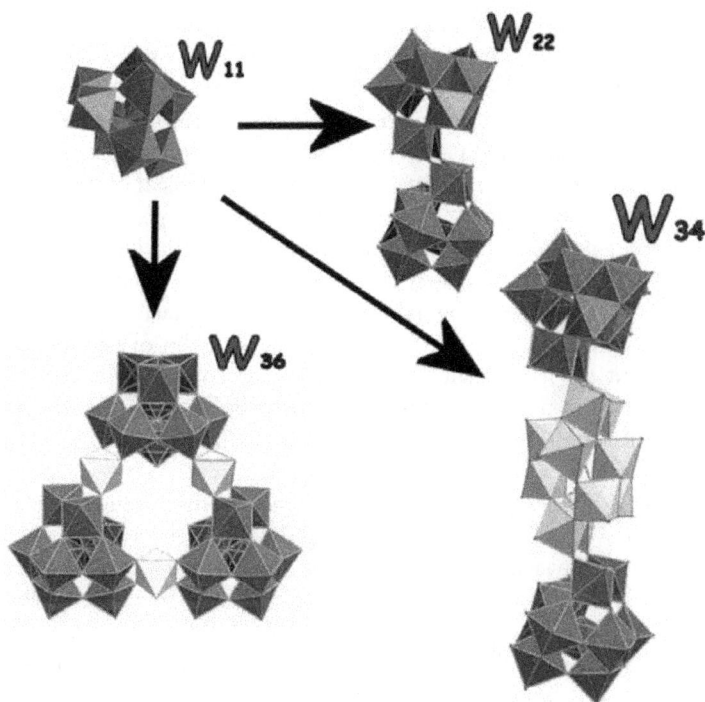

Figure 3. Schema showing the molecular structures of {W$_{11}$} (1 nm) to {W$_{22}$} (2 nm), {W$_{34}$} (3.5 nm) and {W$_{36}$} (3 nm) as controlled pH / counter ions. In this sense the pH and counter ions act as the DNA, causing "translation" of the reaction conditions to give dramatically different structures and nanoscale sizes. The polyhedra show the positions of the tungsten oxide units, whereas the lighter units represent the linkers.

3. How is living technology related to overlapping or nearby research areas, such as nanotechnology, molecular biology, cloning and stem cell research, genetic engineering, and synthetic biology? How is it related to social and technological systems such as social networks or information networks, such as the World Wide Web, cell phone networks, and electronic banking networks?

When living technology is realized, the overlap between it and other areas like molecular biology and nanotechnology will be vast. In fact, I feel that one of the reasons to realize living technology is to shed light on molecular biology and biological nanotechnology. Biological systems are incredibly complex, and the fact they have been assembled according to a global evolutionary process means that understanding these systems from the "top-down" using conventional reductionist research methods is inherently difficult. Conventional cause and effect is complicated by the vast number of possible pathways present in biological systems, so there is a good case to create an "artificial biological system" to abstract the key issues – i.e., to compare artificial genomics with biological genomics, and so on. This is also important if one wishes to "design" functional nanosystems. Reality at the nanoscale is so much different than at the macroscale; motion is fast and furious (due to thermally populated states) and the realm of the uncertain (quantum) is present. This means that conventional design approaches (used to, e.g., build cars or robots) simply will not work at the nanoscale. It could be that the only approach is to use the process of evolutionary trial and error, with the key aspect being focus on the fitness function and the starting points for the system to evolve from, rather than the design of the system.

From my physical sciences point of view, I feel that physical living technologies will be quite far removed from social networks and information networks, at least initially. I think that much of the analogy made between "complex" living systems and "complex" networks is not clearly defined. Perhaps our understanding of the underlying processes that could define "living technology" as opposed to nonliving technology in terms of evolution, fitness functions, system plasticity, robustness, etc. all have merits, but the degree to which there will be useful abstraction will become clear only as a function of the real problems such approaches can solve. Merely using one complex system to help describe and connect with another without any meaningful physical link is rather tough for me to comprehend. Of course, maybe the aesthetics will be important here, since their complex form has a mystical beauty arising from the "wholeness" and the inability to cut the system apart for segmented analysis.

4. What do you think are the most important open research questions about living technology, and how do you think they should be pursued?

As I discussed in Question 1, I think the key question is if a real physical manifestation of living technology will be achieved within the next decade (excluding synthetic biology and AI in computer technology). I think that without this advance, the idea of living technology will not be able to fulfill its potential. Realization of artificial systems will allow us to come to grips with the definition of life and to understand how easy or hard it is for living systems to spontaneously emerge in the universe, and this will also have a multitude of other implications for humankind. Just as important, and possibly even more relevant from a technological point of view, is what impact will living technology have? In this respect, if one subscribes to the ability of living systems to adapt using evolutionary approaches, then the impact on technology could be profound. If the design criteria for any device, drug, material or software could be encoded into a fitness landscape, then the use of a living technology system to evolve toward the product need will profoundly change our world. I think the only viable route to physical living technology will be to demonstrate artificial life-like devices, perhaps even those that would pass a Turing test for life. This needs to be pursued in the chemistry laboratory in both academia and industry, and requires a mix of disciplines from chemistry, biology, physics, computer science, and engineering. Perhaps the main barrier to success in this area lies with the degree of need to get all of the above disciplines to work together at a fundamental level for a significant period of time.

5. What do you consider to be the most interesting and important human or societal implications of research and development in living technology?

The most interesting implication of research in living technology goes beyond the technological consequences, which could be amazing and vast on a scale not matched since the advent of the steam engine and the transistor, but actually tells us something profound about the universe in which we live. For instance, does matter have the intrinsic tendency to evolve to a higher state of complexity or morphological function via evolution, and how possible is it to generate lifelike systems using chemistries that go beyond the relatively narrow chemistry found in biology? These, of course, are important questions, since in our search for extraterrestrial life, we are modeling the signature of life very much on what we have here on Earth.

Acknowledgments

I would like to thank the EPSRC, Leverhulme Trust, Royal Society/Wolfson Foundation, the University of Glasgow and West-CHEM for generous support. I also would like to acknowledge my key collaborators on the EPSRC funded Evolvable Process Design Project – EP/F016360/1 – T. Sun (City University), C. Makatsoris (Brunel University), A. Clark (Warwick University), A. P. Harvey (Newcastle University) and my collaborators on the Chell Project – EP/G026130/1 – B. G. Davis (University of Oxford), C. Alexander and N. Krasnogor (both of the University of Nottingham). I would also like to acknowledge stimulating discussions I had with A. Pross (Ben Gurion University) at Systems Chemistry II, Lake Balaton, Hungary, 18-23 October 2009, which have shaped my understanding of life as a kinetic state of matter.

About the Author: Leroy (Lee) Cronin graduated with a first-class honours degree in Chemistry in 1994 from the University of York, and obtained a DPhil. in bio-inorganic chemistry in 1997 at the University of York under the supervision of Prof. P. H. Walton. After postdoctoral research at Edinburgh University with Neil Robertson and as an Alexander von Humboldt Research Fellow with Prof. A. Müller at the University of Bielefeld in Germany, he returned to the UK as a lecturer at the University of Birmingham in 2000. In 2002 he moved to take up a Lectureship in Glasgow and was promoted to Reader in 2005, Professor in 2006, and was appointed to the Gardiner Chair of Chemistry in April 2009. He holds both an EPSRC Advanced Research Fellowship and a Royal Society-Wolfson Research Merit Award, and is a Fellow of the Royal Society of Edinburgh, Scotland's National Academy of Science and Letters. His research interests including inorganic chemistry, ligand design, self-assembling and self-organizing systems, and he has a deep interest in complex chemical systems. Right now he is trying to assemble and discover inorganic (artificial) biology.

References

Cairns-Smith, A. G. (2008). Chemistry and the missing era of evolution. *Chemistry, A European Journal*, 14(13), 3830.

Cronin, L., Krasnogor, N., Davis, B. G., Alexander, C., Robertson, N., Steinke, J. H. G., Schröder, S. L. M., Khlobystov, A. N., Cooper, G. J. T., Gardner, P. M., Siepmann, P., Whitaker, B. J., & Marsh, D. (2006). The imitation game: A computational chemical approach to recognizing life. *Nature Biotechnology*, 24(10), 1203.

Kindermann, M., Stahl, I., Reimold, M., Pankau, W. M., & von Kiedrowski, G. (2005). Systems chemistry: Kinetic and computational analysis of a nearly exponential organic replicator. *Angewandte Chemie International Edition*, 44(41), 6750.

Miras, H. N., Cooper, G. J. T., Long, D.-L., Bögge, H., Müller, A., Streb, C., & Cronin, L. (2009). Unveiling the transient template in the self assembly of a molecular oxide nanowheel. *Science*, 327, 72.

Miras, H. N., Yan, J., Long, D.-L., & Cronin, L. (2008). Increasing complexity and structural evolution in the assembly of $[H_4W_{22}O_{74}]^{12-}$ 'S' and $[H_{10}W_{34}O_{116}]^{18-}$ '§' shaped isopolyoxotungstate clusters. *Angewandte Chemie International Edition*, 47, 8420.

Pasparakis, G., Krasnogor, N., Cronin, L., Davis, B. G., & Alexander, C. (2010). Controlled polymer synthesis—from biomimicry towards synthetic biology. *Chemical Society Reviews*, 39, 286.

Pross, A. (2005). On the emergence of biological complexity: Life as a kinetic state of matter. *Origins of Life and Evolution of the Biosphere*, 35(2), 151.

6

Ezequiel Di Paolo

Ikerbasque Research Professor

University of the Basque Country

---

1. In what sense do you find it meaningful to talk about "living technology?"

There is a beautiful tension in terms like 'living technology'. Historically, we have tended to contrast the world of machines and the world of living creatures, often seeing them as separate, or even opposed. At the same time we have repeatedly borrowed concepts and metaphors from one world to try to better understand the other. At least since Descartes, animal bodies have been described in mechanistic terms. The metaphors may have changed but this trend has only increased with technological sophistication. We had mechanical, hydraulic and thermodynamic metaphors to depict organic bodies and today we have distributed computer networks as models for complex biochemical regulation, immune reactions, and neural activity. Similarly, we have understood technology in animistic terms, often only heuristically but sometimes with serious consequences. As extensions to the domain of human action, machines are often inspired by natural examples. They can sometimes be regarded as temperamental when they seem keen on pursuing purposes other than our own, making us treat some machines as living beings. The very drive toward automation informs us about the desire for technology to approach the organic and human world as its pinnacle of sophistication and achievement.

These two worlds, that of living organisms and that of human-made artefacts, are also at odds with each other, and herein lies the interesting tension in terms like 'living technology', or similar ones, like 'artificial intelligence', 'artificial life', or 'synthetic biology'. Using machine metaphors to explain life is a useful, but ultimately limited tool for science. And seeking inspiration in organisms to perfect technology sometimes reveals serious contradictions between natural living processes and engineering methodologies. Fundamental disparities between life and (present day) technology become apparent when we consider two essential contrasts between organisms and machines.

Firstly, organisms literally construct each of their components and themselves as integrated wholes in a constant, far from thermodynamic equilibrium flux of matter and energy – this could indeed be taken to be the defining property of life. Anything we consider stable in an organism, whether we are talking about cells, tissues, organs, anatomy, behaviour, knowledge, or social relations, is so only in a precarious sense; things are stable as long as the networked processes of molecular transformation, and several others, keep working. All organic structures are "stable" in a dynamic, non-equilibrium sense, or even unstable but kept within viable bounds, and, as far as we can tell, they all decay eventually. Crucially, this dynamic stability does not rely on material permanence, as is typically the case for the components that make up a machine. In general, each component of a machine remains the same, not by virtue of complex exchanges with other components, but by virtue of its material inertness. We explicitly want machine components to be this way. The resilience of a machine lies largely, though not exclusively, in the resilience of its parts. Machine parts remain machine parts even outside the machine and this cannot strictly be said of organic components. However, far from being a disadvantage, the fact that organic structures are precarious is the root of a different kind of distributed resilience: The autonomy and inherent restlessness of organisms, as well as their flexibility and plasticity, are all properties that we find very hard (nearly impossible) to replicate artificially – all properties, however, that technology envies.

A second and related difference between organisms and machines is that the latter are, without exception, made for satisfying some external goal, while we cannot say the same of organisms. Instead they have an inherently internal purpose: their own continuation as living beings. It makes sense to speak of living organisms as fulfilling an external purpose only when they are used to satisfy human ends (e.g., transport, labour, protection, environmental engineering), in other words, when they are used as machines. Interestingly, our relation to animals becomes less obviously goal-oriented as we interact with them in terms that do not fully deny their autonomy, for instance, in the case of domestic pets. One way to express this contrast is to say that there is a tension between the autonomy of living organisms and the external purposefulness of machines.

Similarly, we find differences moving in the other direction of this relation, from the organic to the technological world. Traditional engineering methods have exploited useful principles of modular design (e.g., divide-and-conquer, combinatorial re-usability, functional compartmentalization). By contrast, "natural design" is often messy, opportunistic, softly modular, functionally distributed, and emergent. It does not have watertight divisions between structures and operates by a (poorly understood) principle of emerging robustness of the whole in the face of unreliable parts, processes, and external influences.

Disciplines like evolvable hardware (i.e., the use of Darwinian principles to evolve functionality in reconfigurable electronic circuits) provide a striking illustration of this difference in how design is approached by nature and by human engineering. In the 1990s, one of the pioneers of this field, Adrian Thompson, from the University of Sussex in the United Kingdom, found something intriguing when trying to evolve electronic chips to do some simple discrimination between different inputs. After many generations, the chips were working as intended, but on studying their evolved structure, Thompson encountered a puzzle: There were components of the chip that were not connected to anything, and yet clamping these components would make the chip not work correctly. How was this possible? To his surprise, Thompson found that these "disconnected" components were in fact interacting through electromagnetic coupling with the rest of the circuit and modulating the processing of input into output in an analogous way. Working directly on the raw material medium allows artificial evolution to bypass constraints that originate in habitual engineering practices, like in this case, that of restricting interaction between chip components to digital information processing and logical gating. Nature, in contrast, doesn't design with a drawing board.

I think it is this double tension that makes 'living technology' an exciting umbrella term. A popular myth about science is that formal deduction is its central driving tool, that scientists extrapolate what the established laws predict and perform experiments to verify them. In fact the driving forces of science are tensions and paradoxes like the ones between organisms and machines. They act as "epistemic gradients" that seek a resolution and fuel creative ideas. This, in my opinion, is what is being denoted by terms like 'living technology'.

2. How does your research relate to living technology, and why were you initially drawn to do this work?

My work explores the continuities between life and mind. I'm interested in understanding the genuine autonomy of living systems, what makes them active and animated, and what makes them build and follow their own purposes and norms and connect to other life. I see life as already mindful in a general sense, even in "simple" cases like bacteria. I'm also interested in understanding the fundamental organizational principles of life, mind and social interaction so that they could eventually be implemented in synthetic systems. So far, the majority of work in cognitive science, artificial intelligence, robotics and artificial life is still framed in traditional, Cartesian assumptions about the mind being something like the problem solving done by a computer inside the head. Such problems are given to the cognitive system externally (by the environment or others) or artificially imposed by the designer (say, a computational module instantiating some form of value system; a box labelled "Motivation" for instance). These designs are very unlike organisms. The latter instantiate an emergent systemic level and, at this level, they actively distinguish themselves as wholes in relation to their environment. It is only at this level, and not at others below it, like that of internal mechanisms, where the language of cognition applies – terms like 'motivation', 'affection', 'decision making', 'behaviour', 'emotion', 'memory', 'learning', 'experience', and so on are valid only at the level of the whole organism. It is simply a category mistake to think that these things happen inside the head, for instance in the brain.

This category mistake, however, is pervasive and is partly due to the lack of a workable, scientific concept of the organism. Biology, psychology, and neuroscience are more comfortable at levels above or below the individual organism, explaining physiological facts in mechanistic terms and general normativity (why organisms do what they do) in evolutionary terms. This is understandable because we tend to use machines as our best metaphor for studying organisms. We approach these questions from a human engineering perspective, but as I said above, organic design works using different, often counterintuitive, principles. My interest in evolutionary robotics, dynamical systems modelling, embodied, enactive cognition and phenomenology is both a desire to explore these natural principles of organization and at the same time to shake off accumulated preconceptions; a quasi-systematic form of un-learning.

Having come from a background of physics and nuclear engineering, then moving on to evolutionary robotics and biological modelling, I have always been interested in complexity and attracted to that region of problems that are nearly impossible to solve. And the keyword here is of course 'nearly'. My personal preference is to find a proper balance in a research problem. It's boring to work in areas where you feel that a lot has already been done, where there are well-established paradigms and methods that work incrementally even if strong, difficult questions still remain open (e.g., some areas of physics). Precisely because the remaining questions tend to locate themselves at the other extreme of difficulty, I tend to find them boring too. This is the extreme of challenges that are so out of reach that we cannot do much about them, we sincerely don't even know how to formulate them (e.g., understanding consciousness) and they demand a radical revision of our vocabularies, concepts and methodologies. The ideal sweet spot is somewhere in the middle, a place requiring creativity and novel methods and where challenging questions grow, but where there is also a chance of succeeding and innovating.

I see disciplines like artificial life, complexity, evolutionary robotics, synthetic biology, etc. also as technologies of thought, tools for clearing up misconceptions and shaking the foundations of habitual ways of thinking. That's why I'm interested in them.

3. How is living technology related to overlapping or nearby research areas, such as nanotechnology, molecular biology, cloning and stem cell research, genetic engineering and synthetic biology? How is it related to social and technological systems such as social networks or information networks, such as the World Wide Web, cell phone networks and electronic banking networks?

Ideally, living technology, conceived as synthetic approaches to understanding life and lifelike processes, should always be in a dialogue with other scientific and engineering disciplines with overlapping subject matters. A proper dialogue should of course go both ways and for mutual benefit. Sometimes this is not so easy; like most interdisciplinary efforts, it requires patience and good communication to overcome terminological, methodological and even community barriers. But when it works, it works well. The best model, of course, is that of a real collaboration between people at the core of each discipline. In this way, living technology can not only draw inspiration from other disciplines, mostly biological ones, but also inform them by providing new tools for knowledge, novel hypotheses and new techniques.

Artificial life is actually a form of theoretical biology. However, theoretical biologists in the past have resisted taking artificial life seriously (sometimes with justification). Still, over the past decade we have seen the gradual but steady adoption by biologists of modelling techniques that were first developed within artificial life. There is now a thriving field of computational biology that owes something to this exchange. So this is possibly another model to follow, apart from direct collaboration: the development of new techniques that override the limitations of more traditional ones.

The same spread of ideas and techniques happens onto wider socio-technological systems, notably in areas of social networks and pervasive computing. Social networking, information networks and portable devices are already changing the way we think, speak and interact with others. These technologies are already getting beneath our skin, like in the case of brain-computer interfaces. However, we mustn't be too surprised about these developments. They are perhaps more overt, but not quite so radically different from learning to do maths, to play a musical instrument or to acquire a new skill or language. The human body is a technological achievement. Cyborg fantasies and realities fall behind the way that humans are always-already artificial creatures. There's nothing natural about the way we walk, talk, dress, manage our physiology, create our own goals, shape our selves as projects, relate to others or even just act and perceive. Human beings are themselves living technology made second nature.

4. What do you think are the most important open research questions about living technology, and how you think they should be pursued?

I have always found challenging the problems that involve, at any level, some complex *transformation* of organization: from plastic experience-dependent changes in a cognitive system leading to altered behaviours and consequently further plastic changes, to systemic phenomena at evolutionary and ecosystems levels such as niche construction, evolution of group behaviour and the constitution of new units of organization and selection. Something happens that makes things not quite like they were before, some in-homogeneity in time, in the rules of the game that is being played. Several themes like the emergence of new functions, the stability of novel structures, and the influence of the collective and higher levels on micro-dynamics all appear recurrently in problems of this kind. And putting these themes together in a conceptually clear framework or in revealing models is still a serious challenge. It's the kind of problem that before the beginnings of systemic thinking scientists would always try to avoid (because it's not clean enough), and we are still not very comfortable with such problems even now.

In particular, we can think of problems such as how to endow a synthetic agent with genuine emergent autonomy, like that of an animal. I don't see the real challenge as that of building artefacts that would have, say, the intelligence of a cat (not a modest goal at all, by the way). The real challenge is to synthesize artificial systems capable of behaving autonomously like a cat, i.e., of establishing what is good and bad for them on their own behalf and acting accordingly, not because we have installed a chip that tells them that food, shelter and companionship are good things. This is a radical challenge implying an understanding of how values emerge in a system as a consequence of its own precarious, and yet dynamically robust constitution, and in interaction with the world and with others. Such "organismic machines" would have a stake in their world encounters, they could not be simply reset, and their experiences would leave formative traces in their organization. What they do, the experiences they undergo, whom they interact with and how, will shape the kind of systems that they become, but they will always have some freedom in shaping themselves. This is in short the challenge: *to build a free artefact.* The big stumbling block between our current situation and such synthetic autonomous systems is our poor understanding of the links between constitutional and interactional domains in living systems – in other words, how what they are both shapes and is shaped by what they do; how the body shapes the mind and the mind shapes the body.

5. What do you consider to be the most interesting and important human or societal implications of research and development in living technology?

I have said that a real challenge for living technology will be to understand the emergent autonomy of life. However, is autonomy a genuine technological goal? In so far as living technology works also as a tool of knowledge, the answer should definitely be yes.

This follows the idea of knowledge by construction. We can be certain to have gained (at least some) knowledge about something once we can build an instance of it. Traditional tools have failed at fully understanding life and the reasons for this may be deeply rooted in the methodological underpinnings of the established scientific method. Formalisms, the core of scientific explanations, have limitations that natural life seems to transcend (cf. the above example from evolvable hardware). We find it hard to pinpoint the formal foundation of concepts like emergence, circular causation, autonomy, agency and cognition. In his *Critique of Judgement*, Kant already appreciated this fact about life: We can recognize it but reason cannot capture its inherent teleology, its defining features. At most, he claimed, we can use our intuitive grasp of organisms as driven by an internal purpose in a regulative manner, to help us guide our research, but we cannot ground these intuitions rationally. In contrast, recent philosophers like Hans Jonas have overturned Kant's conclusion. If a fact of experience, e.g., the intuitions about the interiority of organisms that we use to recognize life as such, is accessible to us thanks to the fact that we are embodied organisms ourselves – and not merely rational beings – then we must not deny such forms of knowledge, but instead we must extend our current methodologies to better account for them. Constructing autonomous, living artefacts is then a genuine scientific goal, precisely because it may reveal to us new forms of understanding life in all its complexity.

Whether autonomy as such is also a goal to be pursued for other reasons, for example, enhancing our lives, is less clear (it is also not so clear exactly whose lives will be enhanced – genetically modified crops can lead to economic dependence, intensive monocropping in third-world countries and the consequent loss of internal markets). Undoubtedly, the marriage of systems engineering and biological complexity may bring innovative solutions to long-standing problems. It will also bring a lot of hype, maybe future financial bubbles and their explosions. Here we can also expect rationality to sometimes fail. The real advances, sometimes modest but still significant, will often be overshadowed by the flashy, headline-making, but ultimately short of groundbreaking claims of the funding and attention seekers. Artificial intelligence gurus have been promising for over 50 years that machines superseding the cognitive powers of their creators will be among us "within the next five years." Likewise, we can expect similar claims about living technology, and we will have to discern the real advances underneath the hype.

But real advances notwithstanding, we can still ask ourselves whether the quest of synthetic life is driven by practical needs (what's the use of a genuinely autonomous planetary explorer if it decides, autonomously, of course, to stay on Earth because it's safer?) or whether something else is at stake. The ethical concerns around current versions of living technologies, apart from the genuine issues and dilemmas that they raise, also point to a different kind of motivation. The most common moral of stories about living technology, from *Frankenstein* to *Bladerunner*, has been that the child-like but powerful autonomy of the creature inevitably turns against the creator. Despite this resonant, guilt-ridden warning, the promise of mastery over nature has been deeply rooted in Western culture since the Enlightenment. And the dream of creating artificial life has been one of its clearest, not to say most obvious, manifestations. This is an inheritance from modernity that science and technology have not shaken off yet. Among the things to be learned from living technology, in its epistemic use as a tool for new kinds of knowledge, we may find a new understanding of integrative, and still productive, fair and egalitarian ways of relating to nature. Not as masters over it, as capitalism would have it, nor as assimilated components of a super-organic whole, as has been the view of both totalitarian regimes and some environmental movements – these views present us with a static, non-transformative concept of humanity and nature. Instead, we may learn, through novel forms of knowledge, to see nature as the Other in a relation of mutual transformation, ongoing and open-ended. Living technology may serve to redefine both nature and our own human selves.

About the Author: Ezequiel Di Paolo was born in Buenos Aires, Argentina in 1970. He studied Physics at the University of Buenos Aires and Nuclear Engineering at the Instituto Balseiro, University of Cuyo, Argentina. After that he moved to areas of artificial life and evolutionary robotics during his DPhil at the University of Sussex. He is now Ikerbasque Research Professor at the University of the Basque country. He has until recently been Reader in Evolutionary and Adaptive Systems at the University of Sussex and co-director of the Evolutionary and Adaptive Systems (EASy) MSc programme. He remains a visiting member of the Centre for Computational Neuroscience and Robotics (CCNR) and the Centre for Research in Cognitive Science (COGS) at Sussex. He is the author of over 100 peer-reviewed publications and his interests include: adaptive behaviour in natural and artificial systems, biological modelling, evolutionary robotics, embodied cognition, philosophy of mind and philosophy of biology. He is Editor-in-Chief of the journal *Adaptive Behavior*, and a member of the Board of Directors of the International Society of Artificial Life.

7

Martin Hanczyc

Associate Professor

Institute of Physics and Chemistry and the Center for Fundamental Living Technology (FLinT), University of Southern Denmark

1. In what sense do you find it meaningful to talk about "living technology?"

I find the term 'living technology' (LT) conceptually interesting since it combines two slippery terms: 'living' and 'technology'. For purposes of clarity I will define both in very general terms. Living refers to a material or process that possesses inheritable information, a metabolism, a body, and layered on top of these traits the ability to replicate and/or internally process inputs, produce outputs and therefore evolve. Technology is a material or process that is produced intentionally to modify the environment in a purposeful way. A living material or process is naturally scrutinized by its ability to survive and reproduce in a varied and often unpredictable environment. A technology is scrutinized by its usefulness and economy.

On the surface, when I think about these terms together (LT) I immediately think of using life toward an intentional outcome. In this category I could place animals (for work), fermentation microorganisms, and even slave labor. This is opposed to other types of technology that exploit natural physical processes that do not necessary involve life in the mechanism, such as a hammer, boat, or wheel, although this is also debatable. This is how I imagine the meaning behind LT.

As technology becomes more sophisticated we can use it in ever-new ways, most recently exemplified by nuclear energy and the Internet. While the utility of nuclear energy lies in its tremendous power output, it becomes useful only if the process is well controlled, and energy is harvested. On the other hand the utility of the Internet partially resides in its uncontrolled distributed architecture. Therefore technology seems to be evolving in at least two ways: toward higher degrees of control (e.g., also in the fine design of nanoscale devices) and also lower degrees of centralized control – both of which are successful as long as the utility function increases.

How does something living become technology? Most of the modern world has become a productive work force. In this way we are acting ourselves as a technology, both for ourselves and for the future generations. The more our social structure and meaning aesthetics change toward production, the more we become technology. As a clearer example, I like to use the domesticated horse, bred purposefully over thousands of years to be a source of work and transport. This is how something living has been gradually transformed into a technology – but it is still alive!

How does technology become living? When we have a process or material that possesses some of the essential characteristics of life as defined above. The more of these characteristics are present, the more living the technology seems. This is pushed to the extreme by the creation of protocells and artificial cells through chemistry, physics, and engineering – the purposeful design and implementation of a living being. However, as technology becomes more lifelike it may also become less useful, practical or economical. Think about whether you would choose to take a car or a horse to your job. A horse is a living being, designed over hundreds of generations to be used as a form of transport. A horse is not only transport but it is smart. It will not carry you into danger. It can sense its environment and make intelligent decisions for you. It is even eco-friendly! This list of attributes is spot on what proponents of LT talk about in their wish list for a future technology. Do any LT supporters use a horse instead of a car? I doubt it. The car is more convenient, predictable, and when not in present use, ignorable. In comparison, a living system or organism employed as technology may fail on convenience, predictability, ignorability, etc. Therefore, when I think about the utility of LT, I cannot see it. If my TV were made of living or semi-living materials, I would not care to give it water, nutrients and dispose of its waste. I also would not find it pleasurable if my TV were making decisions about what to watch. My TV would surely die of neglect. These scenarios may seem ridiculous, but I use them as tools to bring out the concepts underlying LT. Without such examples and reactions to them I personally feel that everyone has a different idea about what LT actually is. My main concern is that the more lifelike a technology becomes, the more undesirable it is as a technology.

One can envision succession of some of our existing technology, which is produced often without any regard for nature in terms of pollution, with a new type of technology that is alive or at least more lifelike, and sympathetically integrated into nature. Such technologies will be incorporated into existing ecosystems, and will not stand against them. Is this in a way bringing us back to a time before the industrial revolution but with new, more sophisticated technologies?

I feel there is nothing new or special about living technology. It has been a part of human culture since before the beginning of human recorded history (e.g., use of animals and people for work). Almost every kind of technology can be called LT, and it fails as a scientific term to describe something new. I personally do not use the term LT in any of my scientific papers. I feel LT is still ambiguous at best and plain missing the point at worst. Perhaps the development of technologies with lifelike properties should be called something completely different that more accurately captures the essence of the idea. One example could be non-equilibrium technologies. Most technologies and products we find useful are useful because they are at equilibrium. Living processes in contrast avoid equilibrium. So maybe this is what some LT people are taking about, but admittedly 'non-equilibrium technologies' does not sounds so exciting. And of course there are other important aspects besides a non-equilibrium state.

This brings me to my final point for this question. I do think that LT is a meaningful term in one way. It is a good term to use outside of the purely scientific realm. It is a good term to engage the public since everyone has an idea or intuitive feel for what 'living' and 'technology' are. This offers an easily accessible entryway into the discussions presented here. A person can be easily engaged to discuss horses, the Internet and the Terminator. In this way I find the term LT most useful and I do use the term in my non-scientific writings and in my collaboration with artists (Richards, 2010). In art as well as science, a primary goal is the conveyance of an idea to other people. Appropriate vehicles are needed to effectively achieve this. LT in the context of art is one such term. Another may be 'artificial life' – another good apparent juxtaposition of terms.

2. How does your research relate to living technology, and why were you initially drawn to do this work?

I am not drawn to LT in my professional research life, rather it draws me in. To date there are two research centers that bear the name LT: the European Center for Living Technology in Venice and the Center for Fundamental Living Technology in Odense. I have not created or named either of them, but I have been fortunate enough to be hired by both centers even though I am not a proponent of the term LT in science.

What I suppose interested these research centers in me is my work involving protocells and the origin of life. As a student at Penn State I studied under the population geneticist Andy Clark. This was my first research experience and I was lucky enough to do it in a first-class research environment. My experience with Andy's group gave me an appreciation for the evolution of systems in general, Drosophila in specific. I then entered graduate school at Yale University to continue my study of the evolution of populations. After a brief stint in human population genetics, my interest quickly turned to *in vitro* evolution of RNA populations under the guidance of Rob Dorit. We were interested in how simple systems could evolve complexity in the time and scale of a laboratory experiment. Then I worked as a postdoctoral fellow under Jack Szostak at Harvard where I wanted to develop a way to evolve populations of self-assembled structures such as vesicles. Relying heavily on the work of Luigi Luisi, we were able to engineer a reproducible, recursive vesicle growth and division cycle. Work in Jack's group brought me even closer to questions regarding the genesis of biological evolution and the origin of life. Work with ProtoLife in Italy taught me an appreciation for complex systems and developing conceptual and practical tools for navigating high-dimensional spaces. My collaborations with friends in Japan (Ikegami, Sugawara, Toyota) helped me develop my current model of a protocell – a responsive oil droplet that can sense and modify its environment and display chemotaxis. This is one of the first artificial chemical systems that can display chemotaxis, thought to be a trait of living systems. The significance of this type of protocell, not only to the origin of life but also to primitive cognition, is the current focus of my research. Given that I have been long involved with playing in the laboratory with living and nonliving systems capable of replication, evolution, and chemotaxis, I am often invited into the LT fold. My research interests overlap with those of the proponents of LT.

Personally I am fascinated by the nature of life and the distinctiveness and evolution of living systems. The great mystery that captures my imagination is how life originated, what constituted the first cell, and what kind of structure (internal and external) provided the necessary condition for life to arise. My efforts to create a new system with lifelike characteristics are only small attempts to understand the mystery. If one day we succeed in creating life, then this will be a great accomplishment, but life creating life is not as interesting to me as non-life creating life.

3. How is living technology related to overlapping or nearby research areas, such as nanotechnology, molecular biology, cloning and stem cell research, genetic engineering and synthetic biology? How is it related to social and technological systems such as social networks or information networks, such as the World Wide Web, cell phone networks, and electronic banking networks?

LT is related to all the above and more. I find LT to be applicable to almost every type of technology (as for example a car that uses energy from the battery to start the motor and then uses the motion of the motor to recharge the battery and thus avoid equilibrium), and to be a mostly meaningless scientific term; see answer to Question 1. As defined in the answer to Question 1, any research field that produces something that has one or more of the characteristics discussed there could fall under LT. This stretches from nanotech to globaltech. Personally, I am quite open about possible overlaps even if not currently existing, in future development of the research fields.

4. What do you think are the most important open research questions about living technology, and how you think they should be pursued?

I find the current understanding of LT to be vague or too general, with many people having different ideas about it. I think more effort should be made to more clearly define the processes LT is trying to promote. According to a website, LT has the following properties: 'self-assembly, self-organization, metabolism, growth and division, purposeful action, adaptive complexity, evolution, and intelligence'[1] I find this description unsatisfactory, especially since I have already argued that many technologies already exist that have at least some if not all of these properties. Perhaps a deeper understanding would bring about an entirely new, but

---

[1] http://www.science-society-policy.org/living-technology

more accurate, terminology, and hopefully it would be more useful. Perhaps more than one terminology of overlapping spectra would be found. More pointedly and in descending order of importance, the most important research questions are:

1) How can we create a living cell or process from the bottom up? Are we currently using the correct methodology to achieve this goal? Most efforts to date to create an artificial cell from scratch use an engineering approach based on design. However, so far this approach has failed to produce an artificial cell. Other approaches can be used that relax the engineering aspect and instead hope to gain some insight from complexity. Perhaps only by increasing the complexity of an artificial system and leaving it as an open system that can be influenced by external factors, can we truly create something that is living. If we succeed in creating artificial life through complexity, then there is a danger that it will be very difficult to prove the creation of life and demonstrate its reproducibility. This method may then stand at odds with the traditional scientific method. Convincing proof and reproducibility have also been major stumbling blocks throughout artificial cell history (Hanczyc, 2008). But in order to create life, perhaps the engineering method is too restrictive to achieve this goal (Ikegami and Hanczyc, 2009). Coupling of open systems with machine learning is one strategy that has potential to achieve the necessary level of complexity to create an artificial cell (Caschera et al., 2010).

2) What is the utility of adding living characteristics and processes (replication, evolution, movement, etc.) to technology? Perhaps in some sectors, using technology with lifelike qualities would be desirable and useful. What are those sectors? In which sectors are such qualities undesirable? Can we even know the answers to these questions at the present time? Ultimately I think that an assessment of current and future technological problems, primarily from the engineering sector, will help to answer these questions and define worthwhile research projects.

3) Is a living TV really what we want? And what are the societal and ethical implications of creating and using autonomous LT?

4) And what about the Terminator?

5. What do you consider to be the most interesting and important *human or societal implications* of research and development in living technology?

The most important implication, because it is personal as well as societal, is whether or not the meaning of life should be: humans = technology. *Aren't we all living technology?* The second most important implication, related to the first, is whether we can redefine our relationship with nature in a sympathetic way that reduces pollution, waste and suffering through new technology, whether it is LT or something else.

About the Author: Martin Hanczyc is Associate Professor at the Institute of Physics and Chemistry and the Center for Fundamental Living Technology (FLinT) in Denmark. He is also an Honorary Senior Lecturer at the Bartlett School of Architecture, University College London. He received a bachelor's degree in Biology from Pennsylvania State University, a doctorate in Genetics from Yale University and was a postdoctorate fellow under Jack Szostak at Harvard University. He is developing novel synthetic chemical systems based on the properties of living systems.

References

Richards, J. (2010). Prophetic visions of a world of living technology. *New Scientist*, 25 June 2010.

Caschera, F., Gazzola, G., Bedau, M. A., Bosch Moreno, C., Buchanan, A., Cawse, J., Packard, N., & Hanczyc, M. M. (2010). Automated discovery of novel drug formulations using predictive iterated high throughput experimentation. *PLoS ONE*, 5(1). Available online at e8546.doi:10.1371/journal.pone.0008546.

Hanczyc, M. M. (2008). The early history of protocells: The search for the recipe of life. In S. Rasmussen, M. A. Bedau, L. Chen, D. Deamer, D. C. Krakauer, N. H. Packard, & P. F. Stadler (Eds.) *Protocells: Bridging nonliving and living matter* (pp. 3-18).Cambridge, MA: MIT Press.

Ikegami, T. & Hanczyc, M. (2009). The search for a first cell under the Maximalism design principle. *Technoetic Arts Journal*, 7(2), 153-164.

8

Inman Harvey

Senior Lecturer

Computer Science and Artificial Intelligence, University of Sussex

1. In what sense do you find it meaningful to talk about "living technology?"

To be honest, I have never used the phrase. However I see it as yet another take on the artificial life quest over the ages to relate living organisms to machines. In the artificial life course that I teach, I make a point of starting with the historical background of such attempts, starting with Hero of Alexandria in the first century AD. One of my favourite quotations is from Thomas Hobbes in *Leviathan* (1651): "For seeing life is but a motion of Limbs, the beginning whereof is in the principal part within; why may we not say that all Automata (Engines that move themselves by springs and wheeles as doth a watch) have an artificiall life?" Each generation relates such ideas to the technologies of their day, but today the advances in computing coupled with technical advances in biology put us in the right position to make serious advances in a systems-level understanding of how biological machines (and biologically-inspired machines) can operate.

This covers a wide area, which I loosely call artificial life, or evolutionary and adaptive systems. Artificial life has always faced in two different directions: on the one hand, using technological artefacts and ideas, such as computers and computation, to assist and enhance our understanding of living systems; on the other hand, exploiting ideas from biology in order to make better technological artefacts. I see 'living technology' as just the latest phrase covering this second sense of artificial life, and (being in cladistics terms a lumper rather than a splitter) I am not very interested in deciding just what should be included or excluded from such a term.

In the nineteenth century, the intellectual and practical movements that transformed the world were based around industrial technology, steam and steel. In the mid-twentieth century, the conception and development of computing laid the basis for the next transformation of our world. Since the latter half of the twentieth century, I think the equivalent transformative intellectual groundswell has been the move towards understanding, at a systems level, how living things work, and exploiting this to develop interesting and useful artefacts. I hesitate to say that we are laying the foundations, more like digging out the footings for laying them – but I expect that in the current century these developments will come to have a transformative effect comparable to that of computing.

2. How does your research relate to living technology, and why were you initially drawn to do this work?

At a personal level, one main reason I was drawn to these ideas was the accident of the particular era I grew up in, and the fact that artificial life was the intellectual zeitgeist. My own intellectual background started in mathematics, then moved to philosophy, with a side excursion into social anthropology. I then had an extended period outside academic life, running my own business, before deciding to move back into academic life and choose a topic for doctoral research. I had from an early age been interested in artificial intelligence – for instance in my teens building a robot based on W. Grey Walter's "tortoise" based on a *Scientific American* article – and I was casting around for ideas. Continuing the *Scientific American* theme, I had long been a fan of the Mathematical Recreations column by the late, great Martin Gardner, followed by successor columns by Doug Hofstadter (of "Godel, Escher, Bach" fame) and by A. K. Dewdney, and these were a rich source of ideas and speculations. I think this was where I first heard about genetic algorithms. So from these influences it somehow seemed obvious, to me at any rate, that my doctoral research should be in evolutionary robotics, although the topic did not really exist at the time.

When I arrived at Sussex, in 1988, to start a Master's degree followed by a doctorate, I was astonished to find that – despite being one of the two UK centres of Artificial Intelligence (AI; together with Edinburgh) – they were only just then starting to teach artificial neural networks (ANNs) and nobody had heard of genetic algorithms. What we now call, patronisingly, GOFAI for Good Old-Fashioned AI was the dominant force, and biological influence had not really arrived yet. I had already programmed a genetic algorithm on my first computer, a Sinclair ZX80, in 1980 or 1981 (I cannot remember whether this was before or after I upgraded it from 1k RAM to 16k RAM, which cost an extra £99.95). I mention this as encouragement to those outside academic life, that cutting edge ideas can be found and developed outside the lecture room. Indeed, since it usually takes 10 or 20 years for new research ideas to become established enough to be turned into a lecture course, and since in AI paradigm shifts can come every decade, it is near-inevitable that textbooks will contain and propagate the discredited ideas of yesterday. If it is in a textbook you should disbelieve it, is my motto; in my artificial life course I claim to try and propagate the lies of today and tomorrow, rather than the lies of yesterday.

The first artificial life workshop was organised by Chris Langton in Santa Fe in 1987, and inspired by that a small group of us set up an artificial life reading group at Sussex. I attended the next Alife conference, with a Sussex colleague Pedro, in February 1990, and have regularly participated since in these, and particularly the series of European Conferences on Artificial Life that Francisco Varela, with Paul Bourgine, started off in Paris in February 1991. Whereas the US-based Artificial Life movement, starting in Santa Fe and Los Alamos, has been strongly influenced by physicists (links with chaos theory, and cellular automata as abstract models of life, for example), the European movement has always had its foundations in biology and philosophy; the influence of Varela is still strong. The influences I have felt most strongly have come from the early European cyberneticians (e.g,. W. Ross Ashby and W. Grey Walter, both with connections to my home city of Bristol); the ANN movement that had similar origins; the dynamical systems approach to cognition that I trace mostly to Ashby; the shift away from GOFAI to New AI in which Rod Brooks played a major role; and the philosophical schools associated with Varela. On the biological side, of course, Darwin is the giant who casts the longest shadow.

I have adopted the strategy of trying to pick out research topics at the edge, that seem potentially flaky and speculative; of course many such topics are indeed flaky and worthless, but it is rewarding to pick out those that have genuine potential, and to help bring them into the mainstream. As soon as such an infant is mature enough to have a band of dedicated supporters, and is thriving, it is time to turn to something else that seems flaky but is worth nurturing.

Evolutionary robotics was the first such field for me, applying analogues of natural Darwinian evolution to the evolutionary design of control systems for robots and other cognitive systems. I credit Peter Cariani with the first usage of such a phrase, and some of the basic ideas date back at least to Alan Turing and to Barricelli in the 1950s; but this was just as thought experiments or computational experiments. In my doctoral research I pursued these, but we also went further to produce the first real robotic system that was evolved in the real messy world, the Sussex gantry robot.

On the algorithms side, I developed the SAGA genetic algorithm that still underlies today what I see as the basic strategy for practical, incremental artificial evolution of working systems. Evolution costs time and money, and to evolve each new species from scratch is so inefficient that natural biological evolution abandoned that approach right from the beginning. In the medium to long term, artificial evolution must follow suit. This means taking designs that have been successfully evolved for some limited set of purposes, and adapting that species through further evolution to extend their capabilities to a new set of purposes. Hence the SAGA acronym, for Species Adaptation Genetic Algorithms. This is less a new GA, and more a mindset for exploiting any standard GA that can cope with tasks of increasing complexity. The main insight is that – as opposed to the typical GA problem, taking an initially random population dispersed across the search space and trying to make this genetically converge on a (near) optimal solution – we will almost always be *starting* with a genetically converged population, a species, and trying to adapt it through further evolution to take on new capacities.

I see some of these basic ideas being inevitably at the foundation of all artificial evolution in the future, not just in robotics. But on the robotics side, I have tried to be influential in the process of taking the field of evolutionary robotics from its proto-existence in the late 1980s as a speculative gleam in the eyes of rather few people, to its current status as a widely recognised field with many practitioners. I claim to be the first person with 'evolutionary robotics' in my job title; but once it became a semi-respectable field, it seemed time to move on.

My doctoral work on artificial evolution drew part of its inspiration from studies by Peter Schuster and colleagues on RNA evolution, in particular the role of neutral mutations where there is a very-many-to-one mapping between genotype space and phenotype space; different individuals that are for all practical and functional purposes identical can be the outcome from a multitude of possible different strings of RNA (or DNA, or artificial DNA). In the context of SAGA, this turns out to be a core insight with important implications. My thesis mentions "ridges," and neutral paths through fitness landscapes, but preceded the generic term of 'neutral networks' that has now come to embrace the phenomenon.

I have been interested in bringing these ideas from the RNA world into practical applications in artificial evolution. In a collaboration with Adrian Thompson, we showed the very real existence of such neutral networks in the fitness landscapes for evolvable hardware, evolving circuit designs for FPGAs (Field Programmable Gate Arrays, basically rapidly reconfigurable computer chips). Lionel Barnett also worked as a doctoral researcher with me to make significant advances in a theoretical understanding of the mathematics and dynamics of neutral networks.

In seeking another superficially flaky (but actually very soundly based) field of research to develop and promote, I came across Gaia Theory and the artificial-life-like Daisyworld model. This was developed by Jim Lovelock and colleagues as a means of understanding how homeostasis might effectively take place at a planetary level, with the interactions between living organisms and their environments apparently giving rise to a network of feedbacks that tended to resist perturbation and maintain conditions appropriate for life. With its new age Gaia terminology, and apparent appeals to unconscious altruistic tendencies for different species to collaborate for the common good, this seemed designed to offend any hard-nosed scientist, and in particular evolutionary theorists of the Dawkins variety. This certainly met my criterion for a topic that was apparently flaky; after a few false starts, and finally simplifying the Daisyworld model to something so trivial that I could understand it, I came to the conclusion that the insight is so fundamentally sound, so basic and so simple, that it has a wide applicability.

My own motivations have been going in the very opposite directions from Lovelock and colleagues, who are appealing to biological notions of homeostasis in order to understand geophysiology, and how the climate interacts with life forms (including particularly humans, currently). I am more interested in taking their models, such as Daisyworld, and using them to produce new insights about homeostasis in natural and artificial life forms; in other words, the "living technology" aspects thereof. Being loosely sympathetic with ideas of autopoiesis pioneered by Maturana and Varela, I see homeostasis, or indeed homeorhesis, as core to any understanding of a living system; what is crucial is an ongoing self-maintaining – indeed self-creating – process that acts so as to resist perturbations that threaten its ongoing existence. Autopoiesis is the ultimate homeostasis of an organism's own organisation and unity.

My main contribution here has been this minimalist simplified version of the Daisyworld model, together with the interpretation of this in terms of "rein control." Picking up on a reference to this by Peter Saunders, one can trace the source of rein control to Manfred Clynes, who noted that many (he wanted to suggest all) biological systems that regulate themselves do not do so by conventional control methods of a single control variable adjusted through negative feedback around a set-point. Rather, there will be two opposing control variables normally in balance. When external perturbations have the potential to disturb this in either direction ("too hot" or "too cold"), one or the other of these control variables will resist. But as with the reins of a horse, which can pull but not push, one needs to resist perturbations in both directions. Clynes and Saunders used this idea to understand natural examples of homeostasis, whereas I have been interested to promote this as a core insight for artificial control systems. It turns out that non-linear systems with multiple feedback loops have a natural tendency to arrive at examples of such homeostasis through rein control – a natural (non-Darwinian) selection against positive feedbacks and in favour of negative feedbacks. This is true at an individual level, an ecosystem level, and a planetary level in the real world, and we can exploit these natural tendencies towards homeostasis in designing complex artefacts. The combinatorial explosion is here on our side; more complexity can lead to more robustness.

Some time ago I developed the Microbial Genetic Algorithm as a minimal GA that was easy to code, effective, and that exploited the horizontal gene transmission common particularly amongst bacteria. More recently the development of Metagenomics has highlighted the symbiotic nature of bacteria. Not only are they somewhat careless in swapping genes amongst themselves, they do not seem to worry all that much as to who does what job, as long as collectively it gets done. So I have recently developed the Binomic GA to exploit this symbiotic aspect as well as hor-

izontal gene transmission; in experiments done with doctoral researcher Nick Tomko this is starting to look surprisingly effective,
and opening up new avenues of research.

**3. How is living technology related to overlapping or
nearby research areas, such as nanotechnology, molecular
biology, cloning and stem cell research, genetic engineering and synthetic biology? How is it related to social and
technological systems such as social networks or information networks, such as the World Wide Web, cell phone
networks and electronic banking networks?**

From my perspective, nanotechnology, cloning and stem cell research have no particular relationship at all with artificial life or
living technology. Many aspects of molecular biology can be of
relevance, since to understand how living systems work one ultimately needs to understand at all different relevant levels – and
the molecular level is, for many purposes, a relevant one. Synthetic
biology is in some sense the mirror-image of living technology: applying engineering principles to designing or modifying biological
life forms, as contrasted with applying biological principles to designing or modifying technological artefacts.

One should appeal to biology when designing technological artefacts that need adaptivity, autonomy, self-repair, and homeostasis;
the crucial and challenging properties of living systems. Robotics,
communication and control systems of all kinds are natural can-
didates.

**4. What do you think are the most important open re-
search questions about living technology, and how you
think they should be pursued?**

Rather than coming out with speculations about widely recognised
current initiatives, I shall focus here on a few initiatives that are
until now largely unrecognised.

Through running a series of Daisyworld workshops with Tim
Lenton I was introduced to the Maximum Entropy Production
Principle. As with Gaia theory, this has been largely in the context of climate models, and it is in this domain that there is the
most convincing experimental support for a controversial idea.
This principle suggests that a system that is in one of many possible non-equilibrium steady-states will tend – subject to whatever constraints are acting on it – to settle in the steady-state
that maximises the rate of entropy production. This is tricky to

nail down, but should be relevant to both thermodynamic and informational forms of entropy. If valid and usable, it should allow us a very general method for predicting the behaviour of living systems; the jury is still out.

The two aspects of my own research, mentioned above, that seem to me to have a vast potential are the concepts of rein control, and binomics. Rein control has the potential to give a fresh perspective on the development of complex living machines, moving away from the traditional negative feedback around setpoints that dominates traditional control theory to the more flexible dynamics of negative feedback through opposing forces. Binomics holds out the promise of having multiple loosely related species of artificial agents cooperating symbiotically – for instance in telecommunication networks – in ways that we are only recently starting to appreciate, through metagenomic studies, as typical in the majority of the natural world. Our views on natural evolution, and thus our versions of artificial evolution, can be changed significantly.

All these are abstract developments that bear fruit in design and in software. The ever-spreading Internet means that autonomous adaptive *software* agents will be the most prominent in the short term – the infrastructure is available almost for free, all that is needed is creativity and insight. In the medium to longer term, we may hope for more material embodied progress: Materials that self-repair, indeed that continually re-create themselves, would be the holy grail that would bring enormous advances going beyond merely software and basic control.

5. What do you consider to be the most interesting and important human or societal implications of research and development in living technology?

Arthur C. Clarke's third law is "Any sufficiently advanced technology is indistinguishable from magic," and the synthesis of artificial life forms, and technologies based around such principles, will certainly qualify. It is inherently near-impossible to predict just which radical disruptive technologies will happen to take off, and how they will affect society, so any speculations are highly uncertain.

The technologies of the past have allowed humans to expand their influence and dominate much of the surface of this planet, increasingly over the last century. In doing so, we generate more and more entropy, and accelerate the onset of thermodynamic heat death. In line with the Maximum Entropy Production Principle, I see this as being furthered by this research. It would be nice to think that the human species will have the typical species-span of a few million years; realistically this looks over-optimistic, though I expect we have a good few thousand years to go. It is going to be an accelerating helter-skelter ride, but with lots of fun on the way as well as disasters. Who knows what life forms will be around after we have left the scene?

About the Author: Inman Harvey's original academic background was in Maths and Moral Science (Philosophy) at Cambridge, with some dabbling in Social Anthropology. Then after many years running his own international trading business, he came to Sussex for doctoral research in Evolutionary Robotics, which did not then exist. This uses skills in Artificial Life, Evolutionary Computing, Genetic Algorithms, and Artificial Neural Networks to synthesise artificial agents by evolving the architecture of their nervous systems and bodies. It can be viewed as Philosophy of Mind using a screwdriver, studying the relationship between behaviours and mechanisms. His philosophical influences include Darwin, Wittgenstein and Varela. He has remained at Sussex since then, as a founding member of the Evolutionary and Adaptive Systems group.

9

Takashi Ikegami

Professor

Department of General System Studies, University of Tokyo

---

1. In what sense do you find it meaningful to talk about "living technology?"

I have recently thought seriously that it is time to bring artificial life (ALife) in silico into the real world. A simple definition of the study of Alife is exploring possible forms of life, irrespective of its constituents' physical and chemical laws and properties. For the last two decades, ALife has shown that there exist many alternative forms of self-replication and evolution in the digital world. But, when I think about technologies around us and wonder how those technologies or some future technologies can be related to artificial life, those concepts proposed by ALife are still very weak. Something is missing.

A major drawback in ALife is that the space-time concept is not well developed. One time step in a computer simulation can be taken as 1 msec or 1000 years in the real world. Also, one space unit can be 1mm or 1km. These space-time scales are scarcely taken into account in ALife. It becomes problematic only when you bring ALife into the real world. Even though ALife seeks possible forms of life, it is worth considering the lower and upper boundaries for life to exist in terms of space-time scales. Life in the real world is all about space and time, i.e., a bacterium and an elephant each have their own space and time organization in their minds. In particular, how to produce a system's own time scales is a critical issue in understanding what life is. To me, living technology is tackling this space-time problem of life to answer the question: What is life? It is not philosophical theory (such as life as negative entropy production), but rather technology that effectively changes our everyday life that will give an answer to the question.

Let us think about the concept of "computation." It is very abstract and hard to imagine, but the concept can be grounded in our physical world in many different ways, e.g., counting by hand, programming by chess pieces, and so on. People may first think the concept of computation too abstract and too useless and may not understand its significance, until a computer has been actually constructed. The same thing is true for the concept of living systems, and living technology. I propose life as "sustainable autonomy;" a system that self-sustains its rich and adaptive dynamics in an open environment.

There exist many abstract and immature concepts around ALife, and living technology means providing ways of realizing those concepts in the real world. Once we succeed in making realization of this concept practical, we will have a better understanding of what life is (and what mind is). All the technology that can be related to ALife is contributing to shaping our ideas of what life is. I thus call those technologies that can help us understand the question (What is life?) "living technology."

2. How does your research relate to living technology, and why were you initially drawn to do this work?

From 1990 to 2004 or 2005, I had worked with ALife in silico, but since 2006, I have switched my research style and decided to bring ALife into the real world. Recently, I conducted three artificial life experiments in the real world; oil droplets, an autonomous robot, and a complex visual machine called MTM ("Mind Time Machine"). As I described above, elaborating the real world implementation of ALife concepts is how I came to be interested in living technology. I shall briefly explain them below:

1) OIL DROPLET

I, along with Martin Hanczyc, Taro Toyota, Naoto Horibe, and Tadashi Sugawara (Hanczyc et al., 2007; see also Hanczyc and Ikegami, 2010), conducted the following chemical experiment: add oleic anhydride oil phase to highly alkaline water phase (pH 12) to see how the hydrolysis of the anhydride proceeds on a glass plate. Immediately the oil began to react with the water, causing the oil phase to break up into smaller spherical droplets, several to hundreds of microns in diameter. These droplets were like gliders (i.e., a soliton-like moving pattern) in the Game of Life, moving freely in the space and interacting with each other. (The Game of Life was originally invented by a mathematician, John H. Conway. It is a simple cellular automaton where a black and white pattern temporally evolves on a two-dimensional grid. Interesting and unexpected behaviour emerges from a simple initial configuration.) The droplets changed direction spontaneously, and they never fused together when coming into contact with one another. In other words, they were far more robust than the gliders in the Game of Life.

Further, the droplets have finite life spans of less than 1 hour and are sensitive to factors in the external environment, such as pH. We argue that the mechanism of the movement is caused by the coupling of the hydrolysis reaction at the interface with the fluid dynamics of the droplet. Because of this coupling, the chemical reaction lasts much longer than without the coupling. This is an artificial life state that sustains the non-equilibrium state in an open environment by its own dynamics.

2) AUTONOMOUS ROBOTS

Julien Hubert, Eiko Matsuda, Eric Silverman, and I (Hubert et al., 2009) made a robot that moves toward a target in its environment. Practically, the robot moves in an open environment where one light source and one sound source are located at the same position but with unexpected noise and light reflections. We studied a simple system by using the Lego Mindstorms NXT, a modular robotics system based upon the NXT microcontroller. Through Hebbian learning between subsystems, this robot contained modifiable connection strengths, so that its structure continually changed during navigating in the environment.

The performance of the robot was measured by its capacity to reach the goal in less than 6 minutes. For a single trial, the performance of a robot would be 1 if it reached the goal and 0 otherwise. We observed that when there existed a certain amount of noise added to the sound sensory inputs, sound and light subsystems become more correlated to each other, which increased overall performance.

3) MTM

Yuta Ogai and I made a machine called the Mind Time Machine ("MTM"), which runs in the real world all day without losing its complex dynamics. (The scientific perspective of this system will be presented at the 12th International Conference on Artificial Life (Odense, Denmark 2010)).

MTM was presented for the first time at the Yamaguchi Center for Arts and Media in March 2010.[1] The machine consists of three screens: right, left, and above, displayed at the corner of a cubic skeleton measuring 5.400 meters per side. Fifteen cameras attached to each pole of the skeleton photograph events that occur in the venue. These images are decomposed into frames, and chaotic neural dynamics control other macro processes that combine, reverse, and superimpose them to make new frames. We presented the MTM as artwork, but at the same time, we recorded data from the system daily to monitor the diversity of the system's behavior.

The operating principle was to process timeframes of the visual inputs by combining chaotic instabilities from neural dynamics and optical feedback in order to make autonomous "time-organizing"' phenomena. Intake images from cameras were progressively embedded into the network's connections as a memory of the patterns. Visual images are taken in and replayed again and again with recursive modifications. The system itself was completely deterministic and used no random numbers, but it showed different images depending on its inherent instabilities, environmental lighting conditions, movement of people coming into the venue, and the system's stored memory.

This was not a large chaotic dynamical system that updates the visual inputs randomly. Different from the mere chaotic system, MTM was designed as a lifelike system, since its dynamics are controlled by an environment and the system has short- and long-term memory to sustain its dynamics. Namely, we claim that MTM is "artificial life," since we designed it to (i) retrieve information from its environment, (ii) memorize it in the form of Hopfield-type learning, which tunes the parameters of the overall dynamics, (iii) generate "episodic memory," (vi) continuously change the network structure via Hebbian dynamics, and (v) organize its overall dynamics as an adaptation to environmental changes.

The three above examples as well as other real world experiments have made me think that it is important to consider the self-organization of space-time scales in the real world. A common feature of the above three examples is that each system has slowly changing processes while interacting with the real environment. I think a system's elaboration of assimilating its behavioral structures into its environment leads to its own space-time orga-

---

[1] See http://doc.ycam.jp/relatedworks/index_en.html.

nization. This behavior assimilation is an example of what living technology means to me, i.e., to design a system's space-time in the real world space-time.

3. How is living technology related to overlapping or nearby research areas, such as nanotechnology, molecular biology, cloning and stem cell research, genetic engineering and synthetic biology? How is it related to social and technological systems such as social networks or information networks, such as the World Wide Web, cell phone networks and electronic banking networks?

When we roughly divide current technologies into living and nonliving technology, my definition of living technology uses the characteristics of living systems. One such characteristic is autonomy, i.e., a system has its own motivation to organize its behavior.

Unfortunately, the current technology does not prefer autonomy, for people tend to think that technology should be subservient to human beings. Autonomous tools that do not meet human desires cannot be good and useful for human beings (e.g., people will never buy an automobile that does not obey a driver's intentions).

On the other hand, there are potentially many studies that require autonomy, not just robot toys. A system should have autonomy where the system must generate a space-time scale that is significantly different from our space-time scale, since we cannot directly apply our design principle to develop technology at those scales. For example, in the field of nanotechnology, we cannot fully control robots as they are in a different space-time scale, so we should let them self-move and improve the situation by themselves.

A vacuum cleaner, "Roomba," is a well-known "autonomous" robot developed by iRobot. Roomba is an example of the point that robot behavior does not have to be fully designed in advance, but should be organized by interacting with the environment. Simple but tedious work like room cleaning certainly needs autonomy, and the Roomba case is just such an example. But, there are other examples where we need autonomy. One can be found where a system becomes so complex that we cannot maintain its processes the way we do with our automobiles. A system must be self-sustainable so as not to lose its autonomous behavior. This is the other facet of autonomy, or let me call it "sustainable autonomy," i.e., a system's mechanism of self-sustainable autonomy; this is what living technology can provide.

Extending this perspective, I think that the World Wide Web also needs living technology to become self-sustainable and autonomous. For example, personal profiles must be self-sustained and autonomously updated. Geographical maps must be self-generated and updated. As we have experienced, the essential problems of an online business are not in setting it up, but in its maintenance. Therefore, autonomy becomes more and more prerequisite for those technologies that must be constantly redeveloped and updated by themselves.

A huge advantage that I expect from those technologies is that a mechanism or algorithm developed to make systems self-sustainable will stimulate life science to answer the question: What is life (and also what is the mind)?

4. What do you think are the most important open research questions about living technology, and how you think they should be pursued?

Living technology is a "living" technology, so it should be tested in the real open-ended environment. Recently, a half-autonomous test satellite, "Hayabusa," received attention in Japan. The mission of this test satellite was to travel 600M km to touchdown on the small asteroid, "Itokawa," and to sample the surface minerals before returning to Earth. Interestingly, the primary mission was not to sample the minerals but to test the efficiency of the ion engine in traveling this long distance and to check how this weak engine could sustain its travel.

Sustainable autonomy must be checked under a real open-ended environment over the long-term. Artificial life has been developed and studied in a "safe" closed environment or examined for a relatively short period of time. When we bring artificial life in silico to the real world, the programmer's fear is losing its sustainability in the open-ended environment over a long period of time. The aforementioned "Hayabusa" satellite was a typical example. It experienced many unexpected accidents that had not occurred in the safe test phases, e.g., 3 out of 4 ion engines stopped, Hayabusa could not touchdown smoothly, it lost the signal, etc. Operators tried hard to reprogram Hayabusa to get it back to Earth, and finally, after 7 years (4 years longer than initially planned), it returned in June 2010. Since the operators rewrote the program from Earth to bring it back, Hayabusa was not a fully autonomous agent. It is not, however, a fully slaving robot either. It can receive signals only after a more than 10 min delay, so it had to compensate for this time delay, showing some partial self-repairing capabilities. What we learned from the Hayabusa story is that the true value of living technology can be tested in longer experiments with seemingly unreachable locations.

Since it is too time- and cost-consuming to examine systems over long periods of time, we need a theory, similar to theory of biological evolution. We cannot yet replicate Darwinian evolution in computers, as it takes so much time. Accelerating evolution in computers, however, is one of the primary missions of ALife studies. The idea of self-sustainability in longer space-time spans must also be replicated and studied. This is a challenging open research question, and its theoretical framework should be pursued seriously.

5. What do you consider to be the most interesting and important human or societal implications of research and development in living technology?

Shusaku Arakawa was an outstanding artist and architect who proposed a new philosophy of life and mind. In particular, Arakawa designed a large space that reverses the destiny of living systems existing in that space. He twisted affordance of the space by designing a new spatial layout to create a new and often radical affordance (e.g., it takes 20 minutes to get from a living room to a toilette, as the floor is rugged and doesn't afford a normal walking style. This unnatural affordance is what Shusaku Arakawa was concerned about). He did this in order to change our unconscious ways of using our body and mind, and thus he insisted our biological life would potentially be extended. He has argued that our mind is strongly shaped by our environment and that our mind is not localized, but distributed in space and time. I had opportunities to talk with him over those issues before he died, and forced myself to reconsider both what the mind is, and the possibility of reconfiguring and recasting living systems with different media, e.g., architecture or some other artificial machines. We don't have to be constrained by the facts of how biological living systems are. Rather we should create a broader umbrella concept of life that includes the concept of biological life, among others.

What does this broader life concept look like? I believe that it is a new kind of technology that enables long-term sustainability, as I have repeatedly mentioned. As a result of such sustainability, it may change our "mind time," reconstructing our mind time in order to have a longer and happier time in our everyday life. A definition of living systems is something that can generate time, and that has a sense of mind "timescape." That is, a living system should organize its own internal time scales in order to let itself be situated in an environment. This idea is the *basso continuo* of MTM and my related studies. In these days, our mind time is compressed and our sense of the momentary "now" becomes narrower and narrower. We need to recover our rich mind time by building a new artificial life machine or architecture. It may be the one Shusaku Arakawa was trying to build in terms of architecture of the reverse destiny. It may be something similar to what I tried by constructing the Mind Time Machine. A living technology of the future may be more "useless" in the sense of current industrial technology, which unidirectionally accelerates our minds. These technologies have been intended to have much more leeway, but paradoxically, we have less time for thinking and talking. What I expect from living technology is that it may fertilize our everyday life by expanding our mind time and cultivating our consciousness. In the end, having such living technologies should provide an alternative way to answer the questions of what life and mind are.

About the Author: Takashi Ikegami received his doctorate in physics from the University of Tokyo in 1989. Currently, he is a professor at the Department of General System Studies, at the University of Tokyo. His research is centered on complex systems and artificial life, a field which aims to build a possible form of life using computer simulations, chemical experiments and robots. Some of these results have been published in *Life Emerges in Motion* (Seido Book Publishers, 2007). Prof. Ikegami frequently attends the International Conference on Artificial Life, and gave the keynote address at the 20th Anniversary of Artificial Life conference in Winchester, UK. He is also a member of the editorial boards of the journals *Artificial Life, Adaptive Behaviors, BioSystems* and *Interaction Studies*.

References

Hanczyc, M. & Ikegami, T. (2010). Chemical basis for minimal cognition. *Artificial Life*, in press.

Hanczyc, M., Toyota,T., Ikegami, T., Packard, N., & Sugawara, T. (2007). Chemistry at the oil-water interface: Self-propelled oil droplets. *Journal of the American Chemical Society*, 129(30), 9386-9391.

Hubert, J., Matsuda, E., Silverman, E., & Ikegami, T. (2009). A robotic approach to understanding robustness. *Proceedings of the 3rd "Mobiligence" Conference* (Awaji, Japan), pp. 361-366.

10

Serge Kernbach

Director

Collective Robotics Group, University of Stuttgart

---

1. In what sense do you find it meaningful to talk about "living technology?"

In robotics, the term 'living technology' is not well established so far, despite the fact that both fields have very similar sounding goals – creating autonomous, self-supporting, sustainable "living" systems. "Living" in this context implies a large spectrum of absolutely different technologies: mechatronics, bio-hybrids, molecular technology, or even virtual creatures in simulation. It seems that different living systems have many common properties; researchers are looking for a deep understanding of these basic principles of what it is to be "living." Concrete examples of living technology illustrate ways in which it is possible to implement these common principles; in other words, technologies represent their embodiment. "Living technology" for this reason may express a common understanding of different facets of natural or artificial life.

Another, more societal aspect of living technology consists in creating new technological beings to accompany the living beings that already exist. To some extent, understanding life means not only being able to create it from scratch, but also improving, supporting, or saving it, or even making it even more advanced. Different socio-economical and socio-technological phenomena can be addressed here. In this way, "living technology" can be thought of as engineering of artificial live and as a science addressing such social aspects as sustainability, stability, safety and other issues.

At the current stage it is not easy to say, which of these two aspects is more relevant. In fact, neither is more relevant – the important element of living technologies is a consolidation of both technological and social elements. It demonstrates in this way a responsibility for technological effects – this is the most meaningful aspect of "living technology."

2. How does your research relate to living technology, and why were you initially drawn to do this work?

Personally, I am fascinated by nature, by its organization, self-regulation, adaptability, reliability, and complexity. Comparing technological developments and their natural "models" – for example locomotive or sensing mechanisms of insects and micro-robots, behavioural strategies, mechanisms of conflict resolution, and energetic efficiency of natural and artificial organisms – we cannot discard the fact that nature is still more efficient, elegant, and reliable. One phenomenon attracted our attention a long time ago – this is the principle of "working together." Two or three or more collectively working elements achieve more results than a simple sum of their efforts without cooperation. This difference, between cooperating and non-cooperating elements, can be an increase in productivity, as in cooperative industrial systems, advanced organization, as in bee colonies, or new structural or functional properties of materials, as in molecular systems. Due to collective phenomena, systems can decrease their level of entropy. Collective phenomena appear everywhere; it seems that a major part of life is based on them.

Robotics is a research area that approaches nature with the principle of "learning by doing." Making creatures that can "think," be self-aware, respond to their environment, possess cognitive features or be self-supporting in energy, contributes most largely to the project of "understanding the origin." Collective robotics inherits all those features of "working together" that so fascinate us in nature – here they are employed in industry, service robotics, surveillance, space and ocean exploration. Modern robotics is no longer confined to electro-mechanical devices, like the robots familiar from old movies. We have now developed micro-mechatronic technology, which involves nano-scale, bio-hybrid robots that combine biotechnology with advanced chip-based, biomolecular and even simpler chemical systems, familiar in modern drug design. As individual robots have become simpler, their cooperation and collaboration has become more important; such robots can succeed in fulfilling targeted tasks only when working in collectives. Here collective phenomena and robotics overlap, in the large research and technological fields of nano-, swarm, networked, and cooperative systems.

Thus, bio-inspired, molecular or bio-hybrid collective robotics, at least from the standpoints of methodology, design principals and functionality, has goals very similar to those of researchers in living technology – i.e., understanding natural and artificial life.

3. How is living technology related to overlapping or nearby research areas, such as nanotechnology, molecular biology, cloning and stem cell research, genetic engineering and synthetic biology? How is it related to social and technological systems such as social networks or information networks, such as the World Wide Web, cell phone networks and electronic banking networks?

From my research background and past experience, I can address only a part of this question. As I said above, bio-engineering, nano-structured materials, micro-mechatronics, and bio-hybrids are converging to a new kind of technology. In robotics, we have a specific term for it – 'disappearing technology'. This means that technology as we now know it – large, independent devices – will become increasingly smart and integrated: in bodies, brains, and the environment. We can observe this trend already today; e.g., existing prostheses are very complex bio-hybrid robots, and embedded microcontrollers can be found almost everywhere: in shavers, power socket, lamps, bicycles, doors and walls, toys, mobile phones and PDAs. In several science fiction visions, humans can directly operate in virtual environments through brain-computer interfacing, or can, e.g., download data directly into their memory. It is not too difficult to imagine many other ways that the micro-world of biotechnologies will merge with the macro-world of information and communication technologies. Living technology can address important issues of convergence of these technologies. The position of living technology, in a broader, more general research landscape, should involve balancing between technological and social aspects, especially in such a sensitive area as medical and pharmaceutical bio-engineering – human beings should remain in focus, and not only technology as such.

4. What do you think are the most important open research questions about living technology, and how you think they should be pursued?

If we consider bio-hybrid, molecular, chemical and nano-robotics as part of living technology, we can identify several open research questions, which are common for different scientific areas contributing to this research field. One of the most important questions relates to the survival and open-ended evolution of these systems in different environments. This involves harvesting energy from available environmental sources, self-repair and self-management, stability and adaptability to various environmental conditions, and principles of self-development. The main open research questions are:

1. What are the driving forces of long-term developmental processes?

2. Are long-term developmental processes controllable? Is evolutionary self-organization controllable?

3. Is there any developmental drift due to emergence of artificial sociality and self-recognition?

4. Are there artificial structural elements that are "absolutely plastic" in the developmental sense, analogous to biological amino acids?

5. Is a "natural chemistry" (i.e., a high complexity of evolutionary processes) important for adaptability and self-development?

6. Is there an "artificial chemistry" that is able to adapt software *in-situ*? Do artificial homeostasis and rules of ecological survival lead to self-identification and to emergence of different self-phenomena (denoted as "self-*"): self-replication, self-development, self-recovering and so on?

The issue of long-term controllability of autonomous artificial systems is extremely important. Artificial adaptive systems with a high degree of plasticity demonstrate developmental drift. There are many reasons for this, like long-term developmental independence and autonomous behavior, emergence of artificial sociality, mechanisms of evolutionary self-organization (which are also a huge challenge), and so on. Such systems are very flexible and adaptive, but they also massively increase their own degree of freedom. New challenges in this area are related to long-term controllability and predictability of "self-*", principles of making plastic purposeful systems, predictability of structural development and goal-oriented self-developing self-organization. These challenges have a great impact on the human community in general (cf. the "Terminator" scenario), as well as in different areas of embodied evolution, like synthetic biology or evolvable and reconfigurable systems and networks.

One of the open research questions is the complexity of "natural chemistry." The development of novel artificial chemistries, for example bio-chemical robots, indicated their ability to re-write the "operating system" in which they are embodied. In biological systems, the chemical machinery of an organism is data and processor simultaneously, thus providing very complex interaction networks but also powerful computation. It is important not only to understand such networks, but also to generate building rules, which allow for building such systems in an engineering way. Basically, developing such chemistries is non-trivial: Many such chemistries at the moment suffer from scaling issues, syntactic and semantic problems and general flexibility. The application of such chemistry raises its own problems, and would indeed also affect the evolvability and stability of the system, since what is being evolved is also in constant flux.

Chemical networks are not the only networks that become complex quite quickly. Interaction networks, which arise in social systems, can also easily get so complex that the cause-and-effect chains are hidden by the overwhelming network of side-effects and indirect causations. Thus studying such systems, and in parallel, developing the tools needed to *understand* these systems are important goals. On the one hand, modern technical networks are still not "autonomous" and "scalable" enough to be satisfying, thus investigating comparably complex interaction networks in nature (e.g., social insect colonies) will provide new mechanisms and novel insights that will help us to understand the emerging complexity in modern-world systems. In parallel, creating such systems from scratch (e.g., in artificial evolution) is also a very promising approach, as long as the products of these "functionality generators" are really investigated and analyzed. Without understanding the "why?" and the "how?" in the evolutionary pathway, these approaches are providing just snapshots, and no generalizable insights.

Finally, the emergence and controllability of "self-*" needs to be mentioned. Different computational processes, leading to global optimization, scalability and reliability of collective systems, create a homeostatic regulation. Homeostasis, as well as artificial hormonal regulation, are important and challenging mechanisms in collective adaptive systems. Moreover, conditions of ecological survival, imposed on such systems, lead to a discrimination between "self" and "non-self," as well as to the emergence of different self-phenomena. There are several great challenges, like understanding these mechanisms or long-term predictability (see above), which have a large impact on the areas of artificial intelligence and intelligent systems, and on the ability to create a new paradigm for adaptive and self-developmental systems. An additional challenge is to be able to "engineer emergence." We envisage systems that are highly evolvable, will adapt themselves over long periods of time, and display emergent properties; today's engineering approaches simply cannot address such a challenge.

5. What do you consider to be the most interesting and important human or societal implications of research and development in living technology?

The implications of living technology in the next five to ten years depend on many factors: general technological developments, positions of policy-makers, breakthroughs in biotechnological fields, level of funding, and efforts undertaken by the community. We can imagine that major developmental steps will be related to a convergence of bio- and micro-mechatronic systems, to disappearing technologies, mentioned above, and to a substantial integration of bio-hybrid technologies into medical and pharmaceutical areas. Each of these three possible scenarios has a large potential impact on individuals and on society in general.

About the Author: Serge Kernbach is the head of the collective robotics group at the University of Stuttgart, Germany. He graduated in electronic engineering and computer science in 1994. During 1996-1998 he received several research grants, and in 2007 his doctoral thesis won the faculty award for the best dissertation of the year. Since 2004 he has been a coordinator of several European research projects in the field of collective robotics. His main research interest is in artificial collective systems. He is the author and co-author of over 100 articles in international journals and conference proceedings, and has edited several books related to robotics.

11

John McCaskill

Professor of Theoretical Biochemistry
University of Bochum

1. In what sense do you find it meaningful to talk about "living technology?"

Concepts are somewhat like children; they go on to have an independent life of their own and their parents remain proud of them, not in spite of but also because of their maturation. While some see "living technology" as a container for a mixed bag of lifelike technologies, I see it as a clearly defined concept, with a strong prototype, for a new type of technology that can radically transform society. While one could include a broad range of biomimetic and bioinspired technology under the term (e.g., using microstructured surfaces, as in the lotus leaf, for self-cleaning tiles), or even technology that interfaces to or otherwise deals with living systems (such as spectacles or harnesses), I have tried to concentrate the strong use of the term on technology that exhibit the *core properties* of living systems that distinguish them from inanimate ones. To put it simply, at least in the minimal physical sense of the word discussed below, "living technology" is technology that is alive.

G. Joyce's definition of life as a "self-sustaining chemical system capable of evolution" may be seen to highlight the three core properties of life: the distinction of self (often referred to as containment), autonomous system sustainment (often referred to as metabolism) and evolution (which since the works of Darwin and Eigen we know physically to depend on genetic – i.e., copyable – information). Of course, any system with these core properties can be termed self-organizing and is likely to have a whole set of other useful properties that we associate with life: self-replication, self-repair, robustness, autopoiesis, adaptation, irritability, etc. Likewise, the advantages of true *living technology* may depend on these more "derived" properties.

Actually, Joyce's definition restricts life to the chemical domain, but none of life's three core properties make this restriction, so that one could replace the adjective "chemical" with "physical," "electronic," or descriptors of other domains without the definition losing its structure. Actually of course, chemical reactions themselves depend primarily on physical electronic structures, and it seems too narrow to restrict life necessarily to a molecular realization. The more general question is whether there is life without a physical embodiment: Actually, many researchers on intelligence also now regard physical embodiment as essential for intelligent autonomous action, so that in fact life may need matter only in the same way intelligence does.

Living technology is technology based on abstraction of the core architecture of life. This is not, as many computer scientists believe, the universal constructor of von Neumann, the rigid architecture of which cannot evolve and does not sustain itself in a complex environment. It does however involve the idea of a self-organized, generic construction capability within a restricted subspace of functional structures.

The interplay between special problem solutions and general ones pervades biology at all levels, from protein translation to the immune system and vision, and is clearly one of the distinguishing features of life. We have shown in two papers how this generalization ability arises in the course of evolution (Füchslin and McCaskill, 2001; Tangen et al., 2006), and it is not a primary property but an emergent property associated with the core features that we have described above. Because of this inbuilt ability to capture generic regularities in the environment under appropriate circumstances, living technology provides a natural route to embodied intelligence, and as such is important beyond the realm of material technology.

2. How does your research relate to living technology, and why were you initially drawn to do this work?

For many years, before formulating the concept of living technology with my colleagues Bedau, Packard and Rasmussen in 2001 at Ghost Ranch, New Mexico, I had been studying living systems in a more physically fundamental manner than traditional in biology. Indeed, I had been interested in understanding the complex dynamics of the chemical systems underpinning life ever since my early research work in Sydney with R. G. Gilbert, but it was only through my time in Göttingen with M. Eigen that I began to delve into the deeper structures in the dynamics of life itself. After resolving some of the key issues with molecular evolution, including showing how the information in quasispecies is limited both for random and more complex landscapes generated by the folding of RNA (these were the first computed molecular fitness landscapes, with complex sequence dependence emerging from physical interactions), I began to investigate more complex functional information processing by molecules using molecular automata models. This led me strongly into contact with information technology for the first time: the computer not just as a tool but as an organization form with strengths and weaknesses compared with life. I simulated many different models of interacting automata, some more abstract, some closer to physical chemistry, and I studied the literature on automata and evolution including such seminal works as von Neumann's self-reproducing automata. This major effort to understand the nature of life, in the broader context of the technically conceivable, pointed my research inevitably (with hindsight) in the direction of living technology.

I might have remained a pure theoretician, quite distant from experiment and all forms of technology, were it not for the strong experimental context in Eigen's lab, which had largely completed the transition from Nobel Prize winning research in fast chemical kinetics to studying physico-chemical self-organization in evolving systems. I realized that spatial inhomogeneities in the distribution of information lay at the heart of Darwin's theory of evolution, and dreamt up a way to investigate the correlation between spatial effects and evolution using the rapidly evolving molecular system based on the Qß enzyme, the technology for which was mastered in Eigen's lab by C. Biebricher and R. Luce. While others in the lab were developing the "Evolution Machine" with extensive machinery, I was designing and investigating spatially resolved evolution of RNA in stationary capillary travelling wave reactors. Together with my young colleague G. Bauer, a talented experimentalist who had mastered the tricky single-molecule contaminant-prone Qß technology, and who later joined the technology firm Evotec, Analytical Technologies, we developed a large area laser-induced fluorescence imaging methodology for parallel evolution in travelling waves. We were able to see the statistical reproducibility of evolutionary processes by measuring tens of thousands of propagating wavefronts: actually all idealized evolution reactors with a fully autonomous equivalent to "serial dilution." Through our study of the "de novo" and macroevolution of sequence information in the laboratory we came to appreciate the creative power of evolutionary processes that can be harnessed by technology.

At this point I began to understand that the technical environment is a crucial component for investigating and understanding living systems. My fitness landscape work then flowered technically in the development of a computational engine for predicting the statistical ensemble of RNA structures, for the first time based on sequence. This challenged the view that a single structure was responsible for function in biology. I was convinced that interactions between proliferating biopolymers would produce interesting functional spatial structures that would teach us about the organization of cells.

So my automata work moved on to focus on interactions, and I developed a vision to build reconfigurable hardware machines that could efficiently simulate interacting molecular automata models for periods equivalent to the history of life on Earth (long time evolution) and for much larger spatial populations than had been possible before. We built the large massively parallel reconfigurable machines NGEN, Polyp and Meregen in the ten years from 1992-2002. At the same time, my experimental work moved on into microfluidics, to provide more controlled spatially structured environments for evolution in which resources could be continually renewed and one could examine long time evolution processes (thousands of generations) under constant conditions. We used both these computations and the suite of microfluidic environments to develop the first in vitro predator-prey and cooperatively coupled evolving systems, based on extensions of enzymatic, isothermal amplification systems for nucleic acid sequences: This was a precursor to modern bottom-up synthetic biology. Both these developments primed my current view of molecular living technology as both intricately linked to evolvability and the interplay between information and the functional capabilities of materials.

In 2001, after an exploration of computation with DNA using magnetooptical reconfigurable microfluidics, I decided that electronic microenvironments provided the best technology platform for extending chemistry towards an evolvable technology. Although many are pursuing a purely chemical and rational design route towards artificial cells, my colleagues Bedau, Rasmussen, Packard and I concluded that what was needed to speed progress was a kind of omega machine: an evolvable chemistry complementation engine that could turn the chemical optimization problem into a coevolutionary process. Although this was too adventurous for the Fraunhofer Society, where a proposal to found the first Institute on Living Technology was rejected in the aftermath of fusion struggles, the European IT community recognized the importance of this development and the PACE[1] project (Programmable Artificial Cell Evolution) was funded by the EU Future Emerging Technologies program. PACE (2004-2008) was successful in building up the framework for living technology in the strong sense, without signing up to deliver artificial cells. Currently, I am coordinating a project to develop an electronic chemical cell (EC-Cell),[2] also funded by the FET Open Program of the EU, and the area of chem./bio-IT has become a proactive initiative at EU level. Living technology has now matured to the point that it is being proposed as the basis of a flagship proposal for IT in Europe.

3. How is living technology related to overlapping or nearby research areas, such as nanotechnology, molecular biology, cloning and stem cell research, genetic engineering and synthetic biology? How is it related to social and technological systems such as social networks or information networks, such as the World Wide Web, cell phone networks and electronic banking networks?

Of course there are issues about the borders of living technology (LT), even when based on the more exclusive definition that I have proposed above. In particular, I would like to recall the distinction between primary and secondary living technology, made in (Bedau et al., 2010), distinguishing LT built from inanimate or animate artefacts. The latter has been around a long time in agriculture, includes the socially controversial area of genetic engineering and is not worth developing a new term for.

Synthetic biology, on the other hand, is obviously more closely related to living technology. Synthetic biology has grown from the modular engineering paradigm often assumed in genetic engineering, based on the half-truths of modular expression of genetic information and the one-way flow of information from genes to proteins and the cell. The recent claim by C. Venter to have created an artificial cell, designed completely on the computer, is perhaps the best-known instance of synthetic biology. Actually, the cell created by Venter and colleagues contained a copy with comparatively minor modifications of an existing cell genome, de-

---

[1] See www.istpace.org.
[2] See www.istpace.org/ECCell.

spite the attention drawn to their watermarking of the genome with personal information. So the claim to have laid the technology for taking us to radically different genomes is premature, and indeed that cell is not yet an artificial cell in the accepted sense of the word.

Other synthetic biology projects are mostly based on extracting other functional subsystems from living cells and modifying or recombining them in new ways. The term 'synthetic biology' is also open to the so-called "bottom-up" approaches, in which novel chemical infrastructure for life is conceived and assembled.

By contrast, living technology proper is less concerned with recombining pieces of existing biological machinery, but rather in harnessing the core properties of living systems in novel ways. It is concerned with taking advantage of making technology more alive, not more lifelike in superficial ways.

Perhaps the closest established area of technology to living technology is evolution technology. Evolutionary computation, evolutionary algorithms and evolutionary optimization now play a significant role in the technical landscape. Evolvable hardware, that has been implemented with reconfigurable hardware, with rapid prototyping equipment (3D printers) and with combinatorial synthesis, is the corresponding embodied wing of evolution technology. Evolution technology is quite advanced at the molecular level, with methods like PCR and Selex now allowing selection of biomolecules as aptamers or catalysts, and recipe-encoding strategies using DNA tags allowing quite advanced information encoding. However, evolution technology has not tackled the self-sustaining nature of living systems. Of course living technology has yet to demonstrate that its technological tool-set, including dynamical systems theory, self-assembly and self-organization theory, can provide a general framework for taking advantage of sustainability and the organizational closure of living systems.

4. What do you think are the most important open research questions about living technology, and how you think they should be pursued?

There are important questions to be resolved at all levels of living technology. On the more general and theoretical side, the control mechanisms for LT and how it should be programmed and designed need to be addressed. What are the appropriate interfaces for human interaction with LT – should we in the end be talking about ecology management rather than industrial programming?

Although it is clear that a metabolism is important in the autonomy of synthesis for life, it is unclear what the influence of metabolic structure on evolvability is. Another question relates to the use of reconfigurability versus reconstruction in LT design; what is the optimal combination? It seems clear that LT synthesis of complex functional structures should involve on-site functional redeployment in addition to positioning by self-assembly of functions. What are the structural limitations of LT construction? How should LT be interfaced with existing technology such as the electronics industry? What are the ethical concerns with LT and how can they be dealt with? And finally, the overarching question: Will the world be a better place when it is based on LT?

More specifically the appropriate questions depend on the research areas and interests of those in the diverse fields that will be attracted to LT. The chemists and materials scientists should be asking not only what are the most appropriate substances and materials to base LT on, but what is the sustainable network of synthetic processes that can best support them? The design of appropriate physical environments with programmable, combinatorial degrees of freedom, as we are striving to attain electronically in ECCell, presents a large area for future research where physicists and engineers can profitably research. This will pay double dividends in that many of the techniques will find application in human interfacing for medical purposes. The question is primarily one of an appropriate division of labour between external specification and internal specification of structures.

Specifically in connection with the personal fabricator (see below), there are questions as to sustainable power collection, the sustainable enrichment of rare resources, synthetic mechanisms, monitoring and feedback control structures, sustainable containment,

the mechanism of deployment, information management, sustainable local waste management, education and training and many more. Progress will depend on young scientists developing prototypes that capture the imagination in specific product areas, rather than a single step to a universal production technology. Much more can be said, but I think that Aristotle's injunction still holds: that we should not try to fill young people with knowledge like pouring water into a barrel, but rather ignite the flame of enquiry that will sustain their own effort independently.

5. What do you consider to be the most interesting and important human or societal implications of research and

development in living technology?

Living technology is crucial if today's society is to overcome the dislocation between industrial production and deployment. Huge factories consume remote resources, and generate products and waste that impact remote areas. The size of the production loop for our key technologies such as silicon chip production is enormous, extending across the globe. The industrial revolution, based on the consumption of fossil fuels and mass-production, has led society into a spiral of mass-produced artefacts that destroys individuality and is unsustainable. At the same time, our machines, including our computers, have been driven into the technological corner, exhibiting efficiency and reliability in an environmental straightjacket, but quite incapable of the robust adaptive behaviour that we associate with life. There is also a software complexity crisis, associated with the completely programmable functionality that has prompted the community to propose organic computing as a solution.

I propose that the next ten years will see significant strides towards the personal fabricator, the manufacturing pendant to the personal computer, and that this will help resolve many of the problems faced by society today. Actually, if taken up by society as a key initiative, personal fabricators could become reality in the course of the next decade. The idea is not to try to build von Neumann's self-reproducing automata, which capture neither the evolvable nor the self-sustaining core properties of life, but to invest in programmable machinery that can take advantage of self-assembly and adaptive processes with feedback to synthesise functional artefacts from local resources. Instead of economy of scale, there will be economy of development time, of deployment adaptation, and of sustainability. The personal fabricator vision requires an investment in living technology, but it can build on progress in reconfigurable hardware, in 3D printing, in combinatorial synthesis, in self-assembly, in systems chemistry and in evolutionary computation. I think that this technology, as with printing and computer technology itself, will most likely at first be quasi two-dimensional. With a personal fabricator, individuals will be able to functionalize surfaces in their home with intelligent processors of material and information, potentially as simply as redecorating. The construction of intelligent digestors of mixed products (a bit like a compost heap) will provide the artificial metabolism necessary for sustainable local production. In a next step, the technology will go fully three-dimensional, moving existing 3D printing to functional self-assembly design using sustainable local materials.

Of course such a development is not possible without access to and control of the domain of chemical materials. Biotechnology is teaching us to master the complex task of programming our processing of macromolecular information, and the integration drive there is moving us from pipetting robotics to integrated microfluidics. Systems chemistry, which emerged in the course of the PACE project as proposed by G. von Kiedrowski, is now playing a major role in the research landscape of Europe and will contribute towards the chemical underpinning of this technology. Organic electronics, DNA machines, self-assembling supramolecular chemistry, and combinatorial materials science will also play a role. It seems clear that artificial cells, although attractive for future living technology, are not mandatory to achieve these goals.

Neil Gershenfeld's[3] vision of a social role for personal manufacturing makes the social significance of this development clear. He argues that the personalization of IT manufacturing technology will provide an enabling self-sustaining industry to bootstrap locally in developing countries, and be economically viable in developed countries for products with a "market of one." We also think that living technology will open access for high technology to the younger generation, as well as to other sectors of society, and thereby allow creative work at the individual and community level. I also believe that it will see the flowering of artistic activity. Of course this all depends on an improved information infrastructure that can support the design process with world-class human interfaces and knowledge. The development of the Internet is already moving in this direction.

Finally, I would like to comment on the growing gap between rich and poor, between creative and non-creative work in society. Living technology is a mechanism that empowers individuals for creative work and innovation, and can help to bridge this gap, for instance through inventions like personal fabrication. This will help alleviate both the injustice of the gap and the social tensions that threaten society as a result. The key phrase here is sustainable innovation: This is possible in society only if it occurs in a distributed manner, on site and in connection with deployment and testing. This is the hallmark of living technology.

---

[3] See http://fab.cba.mit.edu.

About the Author: John McCaskill was born in Sydney, Australia. He studied Chemistry, Physics and Mathematics at the Universities of Sydney and Oxford and is now head of the Bio-MIP research group at Ruhr-Universität Bochum in Germany. His current work is centred on the interplay of self-organization and evolution in lifelike chemical information systems. His early work developed the statistical mechanical basis of molecular evolution, working with Manfred Eigen, and introduced the experimental study of microscale spatially resolved chemical evolution. He created an ensemble approach to RNA folding and self-assembly, now in wide use, and the first reconfigurable computing hardware to simulate long-term chemical evolution. He and his colleagues built the first microfluidic systems for analysing spatial biomolecular evolution and the first programmable *in vitro* molecular ecosystems. He then designed and implemented an optically programmable DNA computer and electronically programmable biomolecular processor using microsystem technology. This has seeded an international initiative to investigate electronically evolvable artificial chemical cells. He has produced over 70 scientific publications. For the past 15 years McCaskill has coordinated a stream of national and international multidisciplinary research projects. He was a founding director of the European Center for Living Technology in Venice.

References

Bedau, M., McCaskill, J. S., Packard, N., & Rasmussen, S. (2010). Living technology: Exploiting life's principles in technology. *Artificial Life*, 16(1), 89-97.

Füchslin, R. M. & McCaskill, J. S. (2001). Evolutionary self-organization of cell-free genetic coding. *Proceedings of the National Academy of Sciences of the United States of America*, 98(16), 9185-9190.

Tangen, U., Wagler, P. F., Chemnitz, S., Goranovic, G., Maeke, T., & McCaskill, J. S. (2006). An electronically controlled microfluidics approach towards artificial cells. *ComPlexUs*, 3(1-3), 48-57.

12

Norman H. Packard

CEO

ProtoLife, Inc., and

Director

European Centre for Living Technology

---

1. In what sense do you find it meaningful to talk about "living technology?"

The term 'living technology' is meaningful inasmuch as there is a useful distinction between living technology and nonliving technology. I will take an operational definition of "living technology" to be "technology that derives its functionality and usefulness primarily from its living properties," and so to assess the meaningfulness of "living technology," let us first consider some of the most important of the living properties that characterize it:

Generation and growth. Technology is typically produced, which is to say, constructed, by assembling parts piece by piece according to a pre-existing plan. In contrast, living systems generally do not come to exist by such a process of piece-by-piece construction; rather, they produce themselves. They grow. They develop. They have no plan that specifies the finished product as part of the plan. An organism's DNA is sometimes referred to as its "genetic blueprint," but a DNA sequence (even a long one) hardly carries all the information to specify every detail of an organism. The DNA is certainly used in a crucial way when the organism proceeds through its developmental process, but its work can take place only within an extremely large and complex biochemical context, the details of which have first-order effects on the final outcome. The complex biochemical context begins with the complex molecular makeup of a reproducing cell, and proceeds, for higher organisms, to change itself throughout the developmental process.

Thus, life "comes to be" rather than being constructed according to a plan.

Homeostasis. Homeostasis, self-regulation or self-mainenance, was one of the main focuses of cybernetics, which is properly seen as a forerunner of both artificial life and living technology. There are many canonical examples of homeostasis that are nonliving, such as a governor or a thermostat. These examples have the character of a system (i.e., technology) sensing the environment, and reacting to the sensations, in a way that maintains the system's state. The canonical examples are rather simple, in that the sensing is simple, the feedback is simple, and the action taken on the basis of the feedback is simple. Living systems tend to have more complex feedback mechanisms, but moreover, they are intrinsically more complex and dynamic than systems typically governed by a thermostat or a governor. When a living system responds to environmental stimuli in order to maintain its living state, the response is usually more complex because its living state is more complex, and we usually think of the complexity of the response being reflected in information processing or computation that takes place as a part of the response. The computational aspect of life's homeostatic mechanisms may lead us to consider life, even primitive life, as having a form of cognition.

Is there a sharp, principled distinction between the way a living system achieves homeostasis and the rather mechanical way a governor or a thermostat achieve homeostasis? Intuitively, there does seem to be a difference, but it is difficult to make the distinction sharp. One difference is that living systems typically must respond to an extremely broad range of environmental variations, some of which may be novel, in the sense that it has never before been encountered. In his studies of cellular automata as models for living processes, von Neumann constructed his self-reproducing machines to be computation universal and construction universal, following the intuition that life may be implementing arbitrary computations. However, it seems unlikely that all life, even primitive life, is capable of universal computation. (It is hard to envision an amoeba computing π to 1000 digits.)

Evolution. Life is amazing, but evolution is the most amazing property of life. Without attempting a formal definition, we may consider evolution to be any system or process that produces variations that then exist in an environment (typically with other variations), and all variations then undergo a process of selection, so that some variations survive and persist, and others die or become dormant.

A consequence of evolution appears to be that system complexity increases with time. This consequence is sometimes contested, usually because complexity is difficult to define, but we will take it to be intuitively certain that the biosphere has increased in complexity, from a primordial state that was so primitive that it has since been consumed by higher life forms, to the contemporary biosphere, including human life with all its complex technology. It is life's increase in complexity that illuminates its most striking feature: Evolution produces an ongoing stream of innovations. It is an inherently creative process. The production of variations feeds the creative process, but only the filtering of the variations through selection yields life's complexity.

The important role of evolution is clarified by considering a phenomenon often presented in the context of attempting to define life: crystal growth. Crystals grow more or less "automatically" in the sense that they are not constructed piece by piece from a pre-existing plan. Their growth obeys physical laws, and the structure that results (sometimes quite complex) is a result of these laws. But the form of crystals cannot always be unambiguously derived from physical law; a vivid example of this is the infinite variety of snowflakes. There is a sense in which crystal growth is a self-healing process; if the environment has a fluctuation, the crystal generally keeps on growing, and often the perturbation diminishes with time, leaving behind a minor dislocation in the structure. But the growth of regular crystals is missing an important feature of life: Variations are not produced and selected preferentially, and so growing crystals do not evolve. The case of complex crystals such as snowflakes is an interesting one, but there is a sense in which selection plays only a weak role, if any (there is enough space in the air where they form so that their essentially infinite variety can exist without selection pressure that would choose one snowflake over another), so they too lack evolution.

The example of crystal growth prompts us to consider more generally the distinction between growth and self-reproduction. Darwinian evolution is generally regarded as involving a process of members of a population reproducing in kind, with variations introduced blindly during the reproduction process. Here, I consider a generalization of this evolutionary process, where an evolving system may extend itself, not just through components reproducing in kind, but also possibly through growth processes. Recall from above that the essential aspects of evolution are (i) production of new system components, (ii) with variation, (iii) followed by selection, and these may be considered independently of the system's mode of self-propagation (self-reproduction vs. growth). Inclusion of growth as a means of life extending itself is biologically not so far-fetched; mushrooms are an example of an organism that has a significant component, a large, spatially extended mycellium, that propagates primarily through growth.

Besides considering general mechanisms for a living system's self-propagation, I also make one other important generalization regarding variation: In addition to random variations, I admit the possibility of variations being nonrandom, and when humans are involved in producing variations, the variations may even be produced intentionally. For a process of change to be evolutionary, the crucial part of the process that must be present is preferential survival of the variations, what I am calling the selection process. The mode of growth and the nature of the variations are quite naturally generalized, if this aspect is maintained.

To be concrete, consider the World Wide Web. It is growing, rather than replicating components in kind, as organisms reproduce themselves. Most of the growth contains variations, and most of the variations are intentionally created by the human beings that install the new components. Yet it is very natural to view the web as evolving.

In my view, evolution is the primary defining feature of life. In fact, the properties of generation, growth, and homeostasis may be considered as defining features only inasmuch as they are needed to support evolution.

Is "living technology" meaningful? Now, after discussing what is meant by the living properties that we presume living technology to possess, we may consider the question, and answer with a resounding affirmative! The first and foremost reason is that the living properties that characterize it are technologically extremely desirable, but generally not present in contemporary technology. Identifying and focusing on these properties will help attain them. The living properties seem to have some linkage, so it may be very useful to understand relationships between them. Moreover, understanding living properties generally, and understanding how they may be instantiated, may be quite useful to particular efforts to develop living technology.

Is living technology really alive? We have referred to living technology as having "living properties," but there is a basic question of whether the living properties of living technology are sufficient to consider the technology alive. Are we talking about "real life" or about an analogy to life? I hold that living technology is, in fact, really alive, if it has appropriate living properties, and in particular if it is capable of evolving. This is a rather extreme position regarding the meaning of "life" and "living," but I believe that as we learn more about life and living properties through study and development of living technology, regarding it as "real life" will become ever more natural.

2. How does your research relate to living technology, and why were you initially drawn to do this work?

I was first interested in studying chaos, and one of the reasons was that chaotic systems may be seen as information generators, in a specific, quantifiable way. Chaos produces random bits, and their randomness is quantifiable with Shannon's and Kolmogorov's information theory.

Evolving systems have always stood out as an example of a different, more interesting kind of information generation. Information is not simply contained in bits, but also in dynamic relationships of components, relationships that implicitly define functionality within an evolving system. Evolution's innovations are the emergence of new functionality. Understanding this aspect of living systems has been the primary long-term goal of my research for quite some time.

3. How is living technology related to overlapping or nearby research areas, such as nanotechnology, molecular biology, cloning and stem cell research, genetic engineering and synthetic biology? How is it related to social and technological systems such as social networks or information networks, such as the world wide web, cell phone networks and electronic banking networks?

An answer to this question depends on the breadth of one's view of living technology. There is a range of living technology, from weak living technology having just a few living properties, or having functionality that depends only weakly on living properties, to strong living technology, which has very strong living properties, and functionality that is crucially related to those properties. But another dimension to consider is the source of the living properties. Some technology might be made of living components, or may itself be alive. An extreme example is provided by synthetic biology, where the technological artifacts are themselves living cells. Such technology might be termed weak living technology not because of the weakness of its living properties, but the weak dependence of those properties on the process of creating the technology. The living properties preceded the creation of the technology. The strongest form of living technology is living (i.e., has a robust spectrum of living properties), but made up of nonliving components.

Thus, the genetically engineered cells of synthetic biology, novel use of stem cells, various forms of cloning may all be considered as forms of living technology. The usefulness of regarding these as living technology will depend on whether challenges and problems of living technology may have common approaches or solutions also relevant to these forms (see the discussion of research challenges below).

Information technology has an interesting property that sets it apart from traditional technology: Much of it is not constructed out of matter. It is informational organization, through, e.g., programming. It has an explicit independence from matter: Computers are theoretically, and to a large extent practically, universal. Software running on one machine is readily ported to other machines. Cyberspace is where information technology resides, and where human use of the technology can realize new forms of communication that lead to new forms of community, which, in turn, can begin to have living properties, and exist as a new instantiation of living technology. For example, as social networks are created, they acquire "a life of their own," in the sense that the social network itself has lifelike properties. In general, all forms of living technology in cyberspace are intermediate, in the sense that some of the components that make them up are themselves living (the people). But the World Wide Web, and the various interacting communities supported by it, have relatively weak dependence on individual human components; their life is a collective phenomenon on a larger scale.

New forms of living technology based on information and communication technology (ICT) are hybrid, in that they involve human beings, as components or participants, in a very strong way. Their evolution is naturally guided and strongly affected by human intention and human psychology. But it is not completely determined by individual human attention and psychology; collective effects may emerge as a result of interactions, without being attributable to individuals.

The future of living technology holds many possibilities that are unforeseeable, because of the emergent nature of the evolutionary process. But there are two broad directions that will almost certainly emerge:

Interacting artificial intelligences (AI's). When William Gibson coined the term "cyberspace" in his novels, he envisioned its population with AI's that had intelligence comparable to or greater then humans, and that were strongly independent of human participation. We have yet to see his vision realized, but we do see software creations that, while they do not have intelligence to rival human intelligence, have enough intelligence to exist and propagate themselves in ways that are largely independent of human beings. Some simple examples of such programs are computer viruses. More generally, production of increasingly intelligent software agents is increasingly plausible, and even probable. Eventually, they will begin interacting between each other, and we will begin to see the emergence of Gibson's vision.

The technological creation of smarter-than-human intelligence has been identified as "The Singularity," and represents the full realization of Gibson's vision. Populations of AI's, interacting both with each other and with the human population, will embody a new level of living technology.

Personal Fabricators. As information technology becomes integrated with fabrication technology, and as this integration proceeds to micro- and nano-scales, we will see the emergence of a form of living technology that may be termed the personal fabricator (PF) in analogy with the personal computer (PC). It will be personal, in the sense that its size and cost will make it accessible to much the same user base as the PC. We presume it will be living technology, because current paths toward its realization are headed in that direction, mostly because the flexibility of living technology may be best suited to universal fabrication. The advent of the PF will have at least as big an impact as the advent of the PC. Just as the World Wide Web has enabled a flowering of new social interaction, so will an analogous web of fabrication. This new web will be driven by human psychology, as is the old Web, but it will have an enhanced capability for expression of human creativity, with instantiations in the material world as well as in cyberspace.

4. What do you think are the most important open research questions about living technology, and how you think they should be pursued?

How may emergence be engineered? Living systems are inherently emergent, in that they are composed of many components that have highly nonlinear interactions, and so they produce phenomena that are unpredictable, and unknowable without observing the emergent phenomena themselves. This is related to the discussion above regarding generation and growth of living systems, as opposed to their construction from a plan. The strong role of emergence makes living systems very difficult to engineer using traditional engineering methods, which generally rely on construction from a plan. Since there is no plan, systems with strongly emergent behavior must be engineered indirectly. Two approaches have proved useful: (i) learning algorithms based on stochastic exploration; genetic algorithms are an example, and recently more sophisticated variants are enhanced by modeling during the learning process; and (ii) judicious application of constraints, to guide emergent behavior toward desired functionality. These approaches are just a start, however, and we are only at the beginning of an understanding of emergent properties of living systems and how we may shape them.

How may evolution be harnessed? Related to the issue of engineering emergence is the question of how the evolutionary properties of living technology might best be put to use. As discussed above, one of the most amazing features of living systems is their ability to create and innovate, in an ongoing way. This property is largely ignored in existing living technology (e.g., the World Wide Web), even though the fruits of the system's evolution are well appreciated.

How may living technology be made safe? As living technology becomes increasingly powerful, it will inevitably have the potential to be extremely dangerous. There are several forms of danger, ranging from the living technology itself running out of control, to human use of living technology, e.g., the possibility of using its capability to fabricate novel substances, to produce extremely dangerous substances. It is clear that as living technology is developed, co-development of safety mechanisms, both technological and socio-legal, will be of paramount importance.

5. What do you consider to be the most interesting and important human or societal implications of research and

development in living technology?

Any technology that provides dramatically increased functionality inevitably carries with it social change. Information and communication technology has already begun social change, through its provision of new modes of human interaction. Other forms of living technology will continue a process of social change in the same vein.

Any powerful technology also carries with it danger. Unfortunately, anticipation of all dangers is inherently impossible for living technology, because of its emergent nature. We must, therefore, develop a dynamic approach to coping with issues as they arise, doing our best to anticipate effects at the earliest opportunity. The emergent nature of living technology means it is impossible to predict everything; it does not mean that it is impossible to predict anything. We must develop social and scientific institutions, norms, and standards that are devoted to anticipating the effects of living technology as it is developed, and that can suggest modes of development, including legal constraints, to safeguard society. We must expect disasters, when we fail to anticipate, or fail to react appropriately, but we must proceed in spite of the disasters we know await (since human controls and safeguards can never be perfect), because the positive potential far outweighs the negative.

The growth of living technology will lead to a greater understanding and appreciation of life itself, and will clarify the relationship of humans with their living environment. As living technology evolves to a personal level, analogous to the way information and communication technology has evolved, individuals will profit from a greater intimacy with life, through interactions with living technology. Whereas social values associated with life now play a relatively marginal role in formation of social norms (the green parties of the world notwithstanding), diffusion of living technology will raise the level of consciousness toward such values. Through living technology, life will become a stronger source of moral, ethical, and aesthetic value.

About the Author: Norman Packard has worked in the areas of chaos, learning algorithms, predictive modeling of complex time series, statistical analysis of evolution, artificial life, and complex adaptive systems.

He holds a B.A. from Reed College (1976) and Ph.D. in Physics from University of California at Santa Cruz (1983). After post-

docs at IHES (Bures-sur-Yvette) and IAS (Princeton), he joined
the physics department at the University of Illinois, Urbana-Cham-
paign in 1987, where he became an associate professor before leaving to become a co-founder of *Prediction Company* in 1991. There
he served as CEO from 1997 to 2003, then as chairman of the
board of directors until 2005, when the company was bought by
UBS. Packard is currently working in a new scientific and business direction based on development of evolutionary chemistry in
for applications in biotechnology. Long-range applications include
the fabrication of artificial cells from nonliving material, and their
programming for useful functionality.

In 2004, Packard co-founded *ProtoLife S.r.l.*, an Italian company based in Venice, Italy, which applies machine learning techniques to the design of experiments (DoE) for high-throughput experiments in biotechnology. He currently works at the US branch
of the company (*ProtoLife Inc.*) in San Francisco, California. From
2004-2008, Packard was involved in the European Commission's
PACE (Programmable Artificial Cell Evolution) project, developing and applying the techniques of ProtoLife. As part of the PACE
project, Packard participated in the founding of *ECLT* (the European Center for Living Technology), where he has served as
director, and as a member of the Center's Science Board. Packard
has had a long-standing involvement with the *Santa Fe Institute*,
and has served on its External Faculty and its Science Steering
Committee.

13

Jean-Paul Peronard

Assistant Professor

Department of Marketing and Management, University of Southern Denmark

---

1. In what sense do you find it meaningful to talk about "living technology?"

There is something special about life. Many people hold life to be sacred. Whenever they hear of scientists meddling with life, or even creating it as in the case of synthetic life, they become anxious and resistant. The sacred involves power and the demarcation of boundaries. It is set apart from the ordinary, everyday, and profane. It may not necessarily be associated with religion, although most religious artifacts and events are considered sacred. To claim that humans can and shall create artificial life is to cross an important boundary that breaks with the fundamental human perception of life being divine. Culture plays an important role in people's reasoning. It enables humans to order the world in such a way that it makes sense. It sets priorities that guide human actions, thoughts and interpretations. Culture also constrains people, inasmuch as it puts limits on their reasoning by dividing the world into categories, i.e. us and them, logical and illogical, true and false, sacred and profane. However, this demarcation is challenged by the complexity of the physical world, and science together with technology plays an important role in establishing new cultural boundaries, since it opens up new ways of perceiving reality.

From the above, it is clear that the concept of living technology will need to inscribe itself in the cultural web of significance, mainly by challenging the sacred, the way we understand life. However, depending on the definition of "life" we can either exclude or include various technologies as "living." Thus, if we want "living" to include religious elements, such as having a soul or a divine purpose, being self-generating, and the like, then a lot of technology may be excluded. But if we instead choose to focus on some basic features of life, which relate to persistence and organization (such as in stem cells and robotics) it seems reasonable to accept a technology as living when it has some core properties in common with life.

The discussion of what constitutes life is not new, and has been a controversial subject for more than half a century, due to an ever faster and more sophisticated technological development. As early as in the beginning of the 1950s Norbert Wiener identified some general similarities between technology (machine) and life (as in a nervous system), which still are relevant today, namely: (1) the performance of tasks requiring independent capacity, (2) the presence of a sensory system, which puts the technology in contact with the surroundings, both to inform about the environment and the technology's own behavior, and (3) the presence of a feedback mechanism that allows the technology to adapt its future behavior based on past performance - a kind of learning mechanism. The robot vacuum cleaner is an example of a technology that with adjustments could resemble "life," although it needs some further development to meet Wiener's criteria. Today, the robot performs a task independently and has a censoring device that controls its direction whenever it encounters an obstacle on its route. However, it still needs to improve its feedback and learning abilities. Presently, the robot vacuum cleaner is unable to communicate with other robot cleaners or with its surroundings. This has implications for using robot cleaners in, for example, an eldercare context. There is a risk of the robot running into people making them fall, or if several robots are working at the same time they could all end up in the same room. In the former case, the robot should be able to receive information from the environment and move only in areas where they are not a nuisance to staff and residents. In the latter example, it is important that the robots be able to communicate with each other and share tasks among themselves. The robots should be able to map the area to be vacuumed and distribute the workload among themselves, and if faced with obstacles in their path, for example a closed door, they should be able to ask for assistance.

As the robot vacuum cleaner examples illustrate, there are more reasons than enhanced functionality and dynamics of living technology that make it relevant to consider a number of technologies under one hat: Namely the fact that living technology is converging in new more complex technical systems. Other examples of such enhanced technical environments are seen in new pervasive medical devices such as in robotic surgery, Pluto mote (a sensing device to monitor patient health) or other wireless sensor networks that gather data over long periods in a patient's environment (e.g., the Harvard University project called CodeBlue). Although the different living technologies should continue to be studied in their own terms and within their own research fields, there is a need to understand the increasing complexity, which will be the result of the emergence of a new ubiquitous network of living technology that includes people, machines, designed biology, objects, etc. We need to study these systems in their entirety because they are part of the human organizing of the world and therefore more than the sum of the individual technologies.

It is important to ensure a degree of mutual adaptation and development between the technological, social and cultural, in order to achieve the best possible synergy between the human and the physical world. Moreover, simplicity does not follow in the footsteps of such adaptation. It requires understanding the complexity of the new environments that living technology generates, and there are no immediate simple analytical tools for understanding this process. Since living technology implies increased complexity, new theories and analytical tools need to be developed. For that purpose we need a mosaic of different reflections, to be assembled in a lens that includes structure, process, and relations. For example, one can change the *structure* of the way health information is collected in eldercare by implementing an electronic data collection and monitoring system, but if it is not integrated with real *process* changes in work task operations, e.g., procedure to respond to the information gathered, the information system is bound to fail. In addition, *relationships* between healthcare workers and their belief patterns also have an impact on the process of technological change. Thus, in a complex system like eldercare, the employees' dreams, fears, traditions, power, social networks, etc., will affect the adoption and use of living technology.

In order to successfully introduce living technology to, e.g., eldercare organizations one should therefore focus not only on the technical finesse and application or enthusiasm among change agents and technical developers, but also on whether the technique is accepted by relevant social groups who can see its benefits. This means that any technology's function and value is embedded in a larger social, political, and cultural web of significance that makes its use and legitimacy apparent for the healthcare workers. It is too simple to see living technology purely from a technical perspective, or to believe the living technology genesis can be viewed as a result of the effect on economic relations. Poor understanding or ignoring any of the above mentioned concepts and relations in the innovation process would be shortsighted and counterproductive, and thus would complicate the further development, introduction, and use of living technology in an eldercare system as well as unnecessarily scaring people.

2. How does your research relate to living technology, and why were you initially drawn to do this work?

What initially drew me into the study of living technology and especially the topic of robots were my interests in social change processes and in doing something productive with my professional discipline, which focuses on the cultural dimension of marketing. After several years of working with technology in different market contexts, it had become clear to me that development had gradually reached a level where many people could benefit greatly from the technological advances, but many ordinary people still distanced themselves from the new technology. At the same time, I was surprised to learn that even the majority of researchers in my professional field had an overwhelmingly narrow focus on Internet opportunities for marketing of goods and services, and only a vague research interest in how emerging technologies could improve the quality of life in the service industry.

I was especially fascinated by the concept of living technology in relation to robotics and the possibility of improving people's lives through the use of living technology, for instance, cognitive support systems that help confused elderly to find their way home. Furthermore, I was happily surprised to find that the scientific research field was intellectually open and less dominated by a narrow worldview, and had a generally positive attitude that things can be changed for the better. Consequently, I recognized that the concept of living technology matches my intellectual curiosity and professional knowledge, and presents an opportunity to contribute to a social change that will benefit society as a whole. However it was only after receiving a government grant for researching living technology in nursing homes that I had the opportunity to realize this.

The grant was for a research project called IntelliCare that seeks to integrate ubiquitous technical solutions for the elderly people in nursing homes. The aim is to make the elderly more self-reliant and secure as well as creating a better work environment for the nursing staff. The project comprises 14 partners from different sectors working together on a wide range of technological development and cross-cutting research projects. The project and its technology solutions target three areas: (1) freedom and mobility – improving quality of life through greater freedom, (2) information – increasing the security of the elderly through easy access to information and as well as helping the nursing staff to manage health information, and (3) scarce resources – optimizing use of resources in daily life, e.g., through the use of service robots.

Traditionally, eldercare has been a service to citizens characterized as "high-touch, low-tech." That is, a service with a focus on personal attention and nursing and with a relatively low level of technology. However, this is changing; new technical aids in the high-tech category have gradually found their way into eldercare and on the political agenda. These technological innovations are wide-ranging. The robot suit Hybrid Assistive Limb (HAL) supports the physical rehabilitation of older people. The animal robot (Paro^{TM}) provides an emotional stimulation to dementia sufferers. Also various alarm and monitoring systems (GPS) may help to prevent unintended and dangerous situations, for example, memory failure among elderly people. Common to these new technical devices is that they provide support for the elderly based on information collected from the elderly themselves and their surroundings through different types of sensors. In addition to bettering quality of life and providing safety for the elderly, the new technology has additional benefits as it reduces stress and anxiety among caregivers as well as providing better working practices and a safer working environment. Another promising aspect of these new technical aids is that they can alleviate a growing shortage of manpower in the eldercare sector in many countries in the western world. For instance in Denmark this shortage will increase over the next two decades due to a skewed demographic evolution in which there will be more people older than 65 years compared to 25-65. Added to this, there are difficulties in both recruitment and retention of workers in the Danish eldercare sector, due to low wages and unfavorable working conditions.

But there also seem to be problems and challenges associated with these new technical devices. Because technology and technique are loaded with meaning and inscribed in a larger social and cultural context, there will often be conflicting perceptions and attitudes as to what a given technology can do and be used for (e.g., technologies for life-prolonging measures, or cleaning people). There is something paradoxical, as well as equivocal, about living technology, which should not come as a surprise since it deals with fundamental human reasoning and understanding. After all, living technology is a cultural product, and culture can be said to contain paradoxes as pointed out by the sociologist Dominique Bouchet. For example, both stem cell research and development of robotics for the elderly represent at the same time a concretization of human dreams and nightmares.

In my research, I am especially interested in understanding how living technology in the form of robotics is developed, selected, implemented, adapted and used within the eldercare sector now and in the future, and how these processes transform organizations, institutions, and society. It is important to study not only a one-dimensional development of the relationship between economics and technical skill, but also the complex interplay between economics, technology, wishes, dreams, fears, identities, attitudes, etc. My starting point is therefore not technology itself, which is neither the problem nor for that matter the solution; it is the way people assign meaning to this technology on the basis of social, cultural, economic and political interests that calls for attention. It forces us to grapple with key issues in technology that unless we tackle them will diminish the possibility of embracing living technology for the better of humanity at all levels. Furthermore, bringing attention to the significance of living technology and providing relevant research is extremely important for policy makers who need knowledge of how and with what means they should support the diffusion on a local level.

3. How is living technology related to overlapping or nearby research areas, such as nanotechnology, molecular biology, cloning and stem cell research, genetic engineering and synthetic biology? How is it related to social and technological systems such as social networks or information networks, such as the World Wide Web, cell phone networks and electronic banking networks?

As mentioned above, living technology in the form of robotics and other complex and ubiquitous systems is constantly changing and finding its way into human life, and in the process bridging previously unrelated domains. We are already beginning to see the merging of humans and technology augmenting human capabilities and creating intelligent networks of linked environments. Examples of such mergers are an artificial skin that soon will cover the human body, and silicon chips that may support the brain capacity. Another example is technology that restores sight to totally blind patients, by implanting electrodes in the visual cortex of a brain and linking them up to a computer. The result of such research has not yet produced perfect eyesight, but has produced a blurred vision that makes it possible to drive a car slowly around a parking lot. In the area of learning and thinking that used to be the preserve of humans, robots are now capable of performing the complicated task of formulating research hypotheses and designing experiments to test them. In addition, robots in the combinations of toys and technological tools have been shown to have great potential for supporting educators and psychologists in their endeavors to help develop skills in humans. Furthermore, as robotics, computers, and humans are entering into wireless connections they will create medical advantages for people living far from urban centers. They will have the opportunity to receive optimal medical attention as living technologies will be able to monitor their health and adjust medication in real time. Clearly this will be viewed as a welcoming improvement for the part of the population living in rural areas, and over time perhaps even an essential right.

As these examples show, the human condition is enormously improved by the collective and complementarily functions of living technologies in the form of robotics, ubiquitous technology, and sensors that connect objects, machines, and people. But for me this description, however, does not quite capture what is at stake. For someone who is still a relative stranger to a field predominantly populated by physicists, chemists, and engineers, the concept of living technology is more than biochemistry, stem cell research, robotics, etc. It's also about the interplay between technology and culture, for the future benefit of society, ourselves, and our children. This calls for an interdisciplinary research design that is able to capture the complexity associated with living technology: one in which researchers from all areas must work together to tackle emerging issues associated with living technology. Through such a mutual awareness of living technology we can ensure choices made at all levels of society are not only technically possible and economically viable, but also culturally compatible, so we may achieve optimum results from these technologies.

4. What do you think are the most important open re-

search questions about living technology, and how you think they should be pursued?

The overriding problem in the coming years is to understand how we can integrate living technology within culture. As the technology evolves, the need for theories and analytical tools to support the process of concentrating the efforts in relevant research areas is growing. There is an urgent need to structure new products and services and to provide organizational management tools that can absorb the new living technologies; that is, to optimize the value of the technologies among a multitude of actors.

The need for more research in the area is further enhanced as living technologies converge and create even more complex technological systems through which humans, and in my case the elderly, are "forced" to experience reality. It is complex because eldercare consists of individual actors whose actions are not always predictable, and because one actor's actions can change the context for other agents. This experience is not just through the Internet, but also through the many cyborg-techniques, which in these years have been developed as visual indicators of the human body's condition (e.g., HAL, the robot suit mentioned above). Already, the technology seems to be a leap forward, and will soon find its way into areas that used to be exclusively for humans. Intelligent machines (e.g., robots) will thus support human activity and thinking, and some will actually merge with our bodies, e.g., memory capacity or new skin. However, while the speed of development is increasing, human understanding of technology is not. It may have important consequences for technology dissemination and utilization, because both the complexity of technical systems and organizational frameworks may limit the understanding, control and expectations of living technology. Such a lack of knowledge creates tension for both individuals and for organizations when selecting and using living technology either individually or in an organizational context. This may then affect the interaction between human and technology and the resulting use of living technology. This is a cyclical process, which of course does not stop here, but in turn will affect the meaning people attach to living technology.

Failure to comprehend living technology may cause people to see it as something external to the human world. It is nevertheless counterproductive to alienate living technology from the social world. The philosopher Gilbert Simondon illustrated this with his technology-philosophy. According to Simondon the technology is part of culture, although it has its own technical genesis. It is an evolving process, which may seem strangely remote to humans, but as the philosopher, economist and psychoanalyst Cornelius Castoriadis points out, it is at the same time an outcome and product of culture; i.e., the social-historical imaginary.

In other words, the process of understanding living technology is characterized by a variety of factors (e.g., scientific, practical, psychological, sociological, anthropological, ideological, economic, political and geopolitical). All these factors are part of a complex relationship, which has effects at both the individual and social levels. It is therefore important to analyze the relationship between these different factors, not least in relation to the foreign and unfamiliar dimension of robotics. The kinds of questions we should ask are:

How are existing systems of relationships challenged by living technology?

What types of reactions, opinions and behaviors occur as a result of new opportunities, problems and challenges in relation to living technology?

What enablers and which barriers exist in relationships?

Is it possible from a cultural analysis perspective to draw implications for management, marketing, research, pedagogy, etc.?

These are not technical questions. They concern fundamental cultural values and meanings. A lack of clear understanding of these issues may create barriers for implementation of living technology among both individuals and organizations. However, these barriers can be broken, converted or diminish with the right approach, and we will use the answers to the questions mentioned above to provide knowledge of how living technology evolves in a social and cultural context. It is important that we provide ways to integrate values with technology and vice versa so we may avoid anxiety and myths, thereby ensuring more optimal benefits.

5. What do you consider to be the most interesting and important human or societal implications of research and development in living technology?

Although one should always be cautious of making predictions about the future, it seems clear to me that living technology will play a major role in eldercare in the coming years, as it will provide means to diagnose and remedy a variety of age-related diseases, aid the elderly in their daily life, and relieve caregivers in monitoring complex sensing data about individuals' health situations.

Consequently, it will reduce the number of hospital admissions, making better use of health care data, but most importantly provide the individual with greater personal freedom, in terms of increased mobility and independence from caregivers.

In addition to improving lives and creating a new technical reality, living technology is and will be a constant source for human imagination and thought. For me, there is no doubt that research in living technology will keep providing results and inspiration that will help mankind set aims for a better future, by opening our eyes to new and improved opportunities to live a good and fulfilled life. Living technology presents new and unprecedented prospects for the individual, organization, market, and society. It will give many people the opportunity to realize their dreams and give hope to the despairing.

But providing these benefits to humans will require an intense effort in research and development of living technology – and there is still a ways to go before we can see actual results. But this may in fact turn out to be beneficial because we will have the time and ability to ensure a cultural compatibility, through serious analysis of what living technology can and will mean for our society, institutions and people. As I see it, social and ethical issues emanate from the field of living technology. Like other inventions, living technology holds as much good and evil as the people who choose to use it. So while we try to do everything possible to get the most out of the promising potential of living technology, we should concurrently do our utmost to explain the significance and consequences in order to ensure broad social support. To succeed in this we need to support democratic debate backed by well-documented social research at the highest level. A significant part of the supporting information may come from technical expertise, but this should be digested and reinterpreted by the social sciences in order to *critically analyze* the tension between the social, economic, technical and cultural dimensions.

Acknowledgements

I would like to thank Professor Dominique Bouchet for his comments and help in clarifying this manuscript.

About the Author: Jean-Paul de Cros Peronard holds a postdoctoral position in the Department of Marketing and Management, University of Southern Denmark. He received a Bachelor's degree in Business Administration in 1995 and a Master's degree in Business Administration (Marketing) in 1998. In 2006, he received his Ph.D. in Business Administration (Marketing). Peronard's main research interest is how technology, change processes, and culture can enhance our understanding of how consumers, organizations, and society function. He is currently research leader on a project called Intellicare. The project explores the possibility of introducing, adopting and using pervasive technologies (e.g., robotics) in eldercare. It is a four-year project sponsored by the Danish Ministry of Science, Technology and Innovation.

14

Steen Rasmussen

Director

Center for Fundamental Living Technology (FLinT), University of Southern Denmark

---

1. In what sense do you find it meaningful to talk about "living technology?"

I believe it is critical for us to start paying attention to the systemic and increasing lifelike properties of the many manmade systems we usually do not consider to be alive. We need to start relating and contrasting these manmade systems to what we usually consider as being alive, because the distinction between the two will increasingly blur and eventually merge in the coming century. The following steps are representative of this trend:

- During the 19th century we automated mass production in factories.

- During the latter part of the 20th century we automated personal information processing in computers.

- The next technological revolution will likely integrate information processing and production into distributed, programmable and sustainable living machines, organizations and systems, with core functionalities now seen in biological systems.

With this technological revolution converging on the horizon, we need to prepare for what will come, as well as start to reflect on what we want and what we don't want when the revolution hits us.

2. How does your research relate to living technology, and why were you initially drawn to do this work?

To me living technology has an important personal dimension.

As far back as I can remember I have been fascinated by life in all of its variations. My childhood was filled with seashells, bird feathers and petrified sea urchins and as a kid I experienced my rural area transforming into a densely populated and polluted vacation spot, all created by humans which themselves are also part of the living world. This dichotomy between "natural life" and "manmade technology" has since both fascinated and bothered me, and a major part of my professional life can be seen as a search for reconciliation between the two.

As it became time for me to pick a profession I wanted to study human ecology, although I did not know that name at the time. However, such topics were not taught, so I eventually enrolled at the Technical University of Denmark. This gave me some freedom in terms of combining subjects across disciplines, although I could not identify with any of the predefined study directions for engineers.

Fueled by a diffuse frustration from not being taught what I needed to learn, a couple of years into my engineering studies I started to wander and I eventually started to study philosophy in parallel with the natural sciences. Since that time, my engineering studies developed into a study of physics and mathematical modeling as I realized that mathematics is the language of the natural world and that physics is the study of constructing simple testable mathematical models of complex phenomena. However, I survived my studies only because of a few exceptional professors who allowed me to explore my interests in a systemic understanding of the world as well as my fascination with self-organizing processes. My PhD and my postdocs at the Technical University of Denmark and later at Los Alamos National Laboratory were a continuation of such explorations, where I studied self-organization in a variety of systems ranging from abstract mathematical and computational processes, to physicochemical processes and socio-technical systems.

The following years at Los Alamos I designed transportation simulation systems, satellite communication simulation systems, urban growth simulations as well as web-based disaster management systems and infrastructure protection simulations. The implementation of these applications was all based on my systemic perspective and my strong focus on self-organization. These applications together with the intellectual environments at the Santa Fe Institute and Los Alamos allowed me to further cultivate and develop my interest in the creative forces in nature and society. This led to the design of a variety of methods and tools focused on self-organization, including a web-based consensus-building and conflict resolution system and a design for a minimal self-reproducing molecular machine.

After co-organizing together the 2000 Artificial Life conference in Portland, USA, Mark Bedau, John McCaskill, Norman Packard and I met again during the summer of 2001 near Ghost Ranch in Northern New Mexico. During this meeting we coined the term 'living technology'. After 2001 things started to move more rapidly in the direction I felt was the right one for me. Most of my professional activities have since been centered around developing the foundation for living technology, e.g., reflected by the establishment of the Self-Organizing Systems (SOS) Team at Los Alamos in 2002, the European Center for Living Technology (ECLT) in Venice in 2004 (as part of the PACE project) and the Center for Fundamental Living Technology at the University of Southern Denmark in 2007.

3. How is living technology related to overlapping or nearby research areas, such as nanotechnology, molecular biology, cloning and stem cell research, genetic engineering and synthetic biology? How is it related to social and technological systems such as social networks or information networks, such as the World Wide Web, cell phone networks and electronic banking networks?

As we inch closer to an understanding of life as a physical process by *constructing* living processes, we are also starting to assess the technological implications of our increasing ability to engineer systems whose power is based on the core features of life: robustness, adaptation, self-repair, self-assembly and self-replication, distributed intelligence and evolution.

I predict an accelerated movement towards more lifelike, living and intelligent processes as well as an integration of living process across many technologies to form new biology-technology ecologies that also include human institutions. If implemented appropriately, these new systems and organizations can become more in tune with human needs and the natural dynamics on the Earth.

I see this development emerging from a knowledge convergence between a variety of sciences and technologies, which we may group into (i) wet carbon-chemistry-based systems, (ii) computational and robotics based, ICT (information and communications technology)-heavy technological systems, and (iii) human organizations and institutions dominated by culture and human nature.

(i) Wet carbon based systems:

Molecular systems are being assembled from the bottom up to form minimal self-reproducing molecular machines while whole genome engineering is modifying existing life forms. The construction of BioBricks, as building blocks for bioengineering, is another example of the emerging synthetic biology. At the center of this research is a desire to understand life by (re-)constructing the elements of life. Research and manipulations of stem cells at the organismic level opens the door for a deeper understanding and control of the more complex functional differentiation processes in biological organisms. As these technologies mature they will eventually touch the more reductionistic synthetic biology.

Utilizing MEMS (micro-electrical-mechanical systems) technologies and microrobotics in connection with computer-controlled microfluidics enables construction of novel hybrid systems with living properties generated by a direct material interaction between the ICT- and biochemically-based processes. The European Commission sponsored projects PACE (Programmable Artificial Cell Evolution; 2004), ECCell (Electronical Chemical Cell; 2008), MATCHIT (Matrix for Chemical IT; 2010), and COBRA (Coordination of Biological & Chemical IT Research Activities; 2010) are examples of activities that explore and develop these possibilities. The integration between biochemical production and ICT-based information processing is at the center of this research.

To close the loop from mass production and personal information processing, I believe that one of the key achievements for living technology in this century will be a personal fabrication technology, somewhat as articulated by several authors elsewhere as well as in this book, e.g., see McCaskill and Packard. This will be a technology that enables ICT-based construction processes to interact seamlessly with material construction processes.

(ii) Computational and robotics based systems:

The physical infrastructures of our societies include the transportation networks, the power production and distribution networks, the communication networks as well as our industrial and food production. Our organizations, our wellbeing and the economy are based on these infrastructures. Due to the increasing complexities and connectedness of the infrastructure elements, robustness, adaptation, self-repair and local intelligence are becoming increasingly more critical for their performance and control. Their control is ensured by increasingly complex and interconnected ICT-based systems based on software agents. The individual components as well as the network of networks of components need to become more lifelike to be reliable and agile.

Further, the physical systems components also need to become more lifelike out of a need to become sustainable processes in a more balanced global human-natural system. Ultimately this means that our economic system needs to be expanded to include the necessary physical and social externalities in the definition of wealth and financial transactions.

Already today, regional agriculture can be transformed into sustainable ecosystems through ecological farming and the inclusion of new ICT-supported praxis, e.g., in terms of GIS (geographic information system) for precision soil treatment and robotics for milking as seen many places. ICT-based reconfigurable production plants can address the need for costumer-centric demands for different products at different times as discussed elsewhere in this book. Such adaptive industrial complexes can further be reengineered to become more self-referential and ecologically sound with the development of industrial metabolisms, where the waste stream from one plant in part becomes the input of another plant, similar to the material circulation seen in natural systems or ecological farming. Whole regions (e.g., Sønderborg in Denmark, coordinated by ProjectZero) and islands (e.g., Samsø, also in Denmark) are today in the process of becoming CO_2 neutral over the next 20 years based on these possibilities. Looking at these processes the greatest challenges are, in my opinion, not the technologies which are largely there, but us and our institutions.

(iii) Human organizations:
The most critical aspect of many current human organizations and institutions is that they act as roadblocks for the necessary innovations, rather than enablers. Inadequate human culture and traditions as well as some aspects of human nature make many human organizations dysfunctional. This is discussed in depth by other authors elsewhere as well as in this book, e.g., see Ulieru. Again, knowledge about and inspiration from living systems could mitigate such dysfunctionalities, as bottom-up processes could self-organize together with top-down coordination and make institutions more adaptive and human-needs centric.

4. What do you think are the most important open research questions about living technology, and how you think they should be pursued?

Each of the above-mentioned research and technology areas have their own open research questions associated with their domain-specific methods and techniques.

- For the wet carbon-based systems the issue is to explore, create and control physicochemical, mechano-electronic and biological processes such that living processes become clear and operational.

- For the agent-based computational and robotics-based systems, the issues are to make the domain-specific systems more lifelike and intelligent.

- For the human culture dominated systems, the issue is mostly to reorganize into more adaptive, agile, sustainable, and human-needs-centered institutions, which in part can occur by utilizing distributed, living, information and communication technologies.

Externally, I believe the following issues will drive scientific and technological transformation:

- There is a lack of sustainable production of goods and services at all levels, which, e.g., results in global warming.

- The implementation of critical innovations is inhibited due to archaic institutional barriers. E.g., we have hunger at the global scale while we produce more food than we can eat. At the local scale these organizational inefficiencies hugely waste both human and physical resources.

- We are experiencing a complexity crisis at most levels of society, and for most of our technologies. There are fundamental challenges in both the details and the control of top-down engineering of technology, e.g., in software development and the development of efficient and more sustainable infrastructures. There is alienation and disengagement in civic processes as well as in our industrial and public institutions. There is, e.g., a decreasing focus on insight and scientific competency as critical parts of policy making, and inadequate organizational mechanisms for resolving workflow conflicts and waste in industry and the public sector.

- Many problems with current technology and society result from the impact of a dislocation between design, production and deployment. E.g. today producers and consumers are different and often from different continents, which means that consumers usually are unaware of any negative humanitarian or environmental impacts stemming from their consumption.

The good news is that I believe that each of the above mentioned science and technology areas can produce breakthroughs in the coming years, as they all have vibrant emerging communities acutely understanding these issues and seeking to make "their systems" more alive and/or more intelligent. The bad news is that most of our current human institutions (national and regional governance, industry and universities) are inadequate to deal with the consequences of such changes. Our institutions have to become part of this transformation and become more adaptive, agile and innovative.

I'm most intimately involved with the community that seeks to assemble physico-chemically based life from scratch. Minimal life can be defined in this context as a system capable of converting resources into building blocks, enabling the system to grow and divide, which in part needs to be controlled by an informational system, where the information can change from generation to generation. Different instructions for growth and division in different individuals thereby enable selection and thus evolution.

In practice this means that our community needs to couple an informational system and a metabolism within a container; different groups have different detailed designs for how to do this in practice. For our design we have demonstrated how minimal information can act as a catalyst for the metabolic production of containers and how the process depends on the information, how containers can divide as a result of growth and how successive feeding of resources can occur as well as how a balanced growth of information- and container replication will occur when they are coupled in the catalytic manner we are implementing. Our teams at Los Alamos and at the University of Southern Denmark have obtained these results since the early 2000s. For details, e.g., visit http://www.sdu.dk/flint.

Our team's immediate overarching experimental challenges in this context are to demonstrate that an informational system can replicate while attached to the exterior of a container, followed by demonstrating a coupling of this replicating information system with our already working information-controlled metabolism. Even though this can go wrong in a million ways, I believe this is possible in the near future. The main theoretical challenge is in my opinion to demonstrate how information, kinetics and thermo-dynamics can interact to self-organize matter into a living process. I also believe this is an achievable task in the near term.

5. What do you consider to be the most interesting and important human or societal implications of research and development in living technology?

Advances in living technology, which will be based on the devel-opment and the reengineering of natural and artificial living and intelligent processes, including human organizations, will impact virtually all sectors of our societies (see Figure 1).

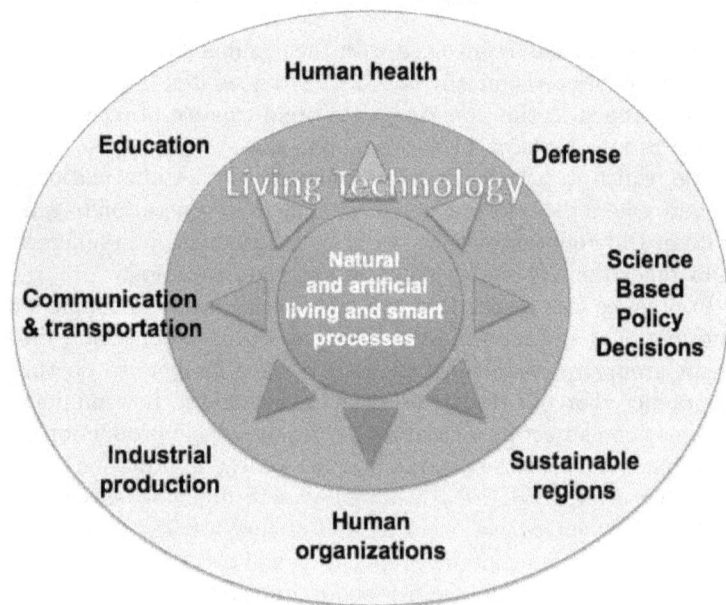

Figure 1. Living technology will impact virtually all sectors of society and at the core these living technologies are based on the implementation of living and intelligent processes at all levels.

I believe that an efficient and practical way to prototype different implementations of the broader living technology vision would be in geographically bounded regions. A significant part of the necessary processes in such a transformation requires deep stakeholder involvement and organizational changes as well as reengineering of legacy infrastructures in addition to the more science- and technology-heavy components. Therefore a region seems to be at the right scale as it would both focus the living technology implementation efforts and solve a variety of pressing societal problems, including the implementation of CO_2 neutrality, which are already well under way, and more agile and human-centric organizations and infrastructures. Due to the nature of the necessary socio-technical and policy processes a certain minimal scale is needed. Also, I don't see such a transformation possible in any direct way at a national or at a global scale. However, successful regional experiments could fuel a societal transformation into more living and livable situations, which are more in tune with human needs and our natural possibilities and limitations.

The underlying vision for developing the above systems is the idea of engineering manmade systems at all levels to become more like living organisms or ecosystems, both to better accommodate human needs and to obtain a more balanceed interaction with the biosphere. This is not my idea, but a vision I see emerging from part of my scientific community (including many authors of this book) as well as part of the progressive business and NGO communities, although different labels for similar things, processes and organizations often are used.

However, the development and the implementation of pervasive living technologies can serve both good and evil, which of course has been true for all human technologies. But because of the powers of living processes, we really need to pay attention in this case. I believe it is paramount that we have an open dialog about what living technology could be and what we want it to become. See, e.g., the discussion by Bedau in this book. In the long term this implies that we'll have to reflect on what it means to be human in a world with both living biological and living technological processes, as well as reflecting on what these new ecosystems could or should evolve into. We should not forget that we have evolved into humans only in very recently geological times and that it is naïve to believe that we, as the humans we are today, are the end product of evolution. The emergence of living technology forces us to be humble and realistic as well as bold and cautious, because

living technology critically challenges who we are as well as what
we may become.

About the Author: Steen Rasmussen (Ph.D Technical University of Denmark, 1985) focuses mainly on pioneering and im-
plementing new approaches, methods, and applications for self-
organizing processes in natural and human made systems. He
is currently the Director for the Center for Fundamental Living
Technology (FLinT), Research Director at the Department for
Physics and Chemistry at University of Southern Denmark (SDU),
External Research Professor at the Santa Fe Institute (SFI), USA,
Principle Investigator (PI) of the European Union (EC) sponsored
Matrix for Chemical IT (MATCHIT) project and Co-PI for the
EC sponsored Electronic Chemical Cell (ECCell) and Coordina-
tion of Biological & Chemical IT Research Activities (COBRA)
projects. He was also the PI for the startup of the Initiative for
Society, and Policy (ISSP) in Denmark, the Team Leader for the
Self-Organizing Systems team at Los Alamos and a Guest Profes-
sor at University of Copenhagen (2004-5). He was PI for the Los
Alamos Protocell Assembly (LDRD-DR) project and the Astrobi-
ology program (origins of life) at Los Alamos, developing experimental and computational protocells and Cell-Like Entities, with
USAF as a co-sponsor. Further, he was the co-director on the European Union sponsored Programmable Artificial Cell Evolution
(PACE) project, and he was one of the founders of the Artificial
Life movement in the late 1980s. He was the Chair of the Science
and Engineering Leadership Team (SELT) for 2001-2002 in the
Earth and Environmental Science (EES) Division at LANL and is
currently on the Science Board for the European Center for Living Technology (ECLT) in Venice, Italy, which he co-founded in
2004. He also currently heads the Science Board for ProjectZero
in Sønderborg, Denmark.

Professor Rasmussen has published more than 85 peer reviewed
papers and many internal technical reports, given more than 175
invited presentations outside of home institutions, and co-organized
eight international and several national conferences. He organized
the first two international protocell meetings, one at Los Alamos
and the Santa Fe Institute (US) and one in Dortmund (Germany),
and edited the first book on the topic. Many communications
about his work inside and outside of the scientific establishment
have appeared on television and in newspapers, periodicals, and
books. Since 2000 he has sponsored 15 postdocs (theorists and experimentalists) and 30 graduate and undergraduate students. He

is also actively engaged in the public debate about science, society and policy.

15

Markus Schmidt

Co-Founder and Project Leader

Organisation for International Dialogue and Conflict Management

---

1. In what sense do you find it meaningful to talk about "living technology?"

In recent years we have heard a number of buzzwords such as nanotechnology, systems and synthetic biology, and converging technology, where it was not always totally clear how these terms were actually defined. Inventing and using new names can be an attempt to demonstrate a fundamental change in the way science and engineering is done, and it can also serve to create a community of people sharing similar interests, including the need to convince policy makers to provide new funding opportunities. Now we see a new kid on the block, branded "living technology." While living technology could be seen as an attempt to mobilize new funding opportunities with a new catch phrase, I prefer to see it as the starting point for the construction of smart and renewable matter.

So what is living technology? With a little help from the *Encyclopædia Britannica*, I define living technology as the application of scientific knowledge to the practical aims of human life, using autopoietic matter that is characterized by the ability to metabolize nutrients, grow, reproduce, and respond and adapt to environmental stimuli. In bioengineering, life (as we know it) and its subsystems are used as the primary resource for doing engineering. Existing systems are manipulated and engineered in order to yield useful results, something we may refer to as wetware, in contrast to hardware (e.g., robots, electronic circuits, and other engineered non-living matter) and software (computational processes, manipulating electrons on a chip, etc.). In contrast to bioengineering, living technology aims at engineering matter to have characteristics of the living world using all three "wares:" soft-, hard- and wetware.

In a way living technology is a kind of converging technology, as wet-, soft-, and hardware are combined; however, the difference is that living technology always has lifelike characteristics, while this is not the case for converging technologies (a term describing the convergence of nanotechnology, biotechnology, information technology, cognitive science and robotics). While living technology intends to produce engineered systems that have lifelike properties, they are different from naturally evolved life forms. In a way living technology tries to release life (in its abstract form) from being "locked" into naturally evolved cells and its specific set of biomolecules. There is no doubt that this is incredible difficult, as living processes are highly complex systems; however, while life is highly complex, it is not infinitely complex. This means that life and its complex processes can in principle be understood and put into engineering systems, although this has not yet been realized. For living technology to be successful, we need to be able to engineer a technological system that is complex enough to exhibit characteristics of living matter, using wet-, hard- and software. In the long-term future we might even find a way to construct other or more complex systems that exhibit new functions that current life forms do not exhibit.

To sum it up, I think it is meaningful to talk about living technology because there is no established term available to define engineered matter that consists of wet-, soft-, and hardware and exhibits lifelike properties. My guess is that it will be useful to talk about living technology until we reach yet another level of complexity in research and engineering where we find characteristics unknown to and beyond life as we know it.

2. How does your research relate to living technology, and why were you initially drawn to do this work?

I started my professional education with a five-year programme in electronic and biomedical engineering, where I learned to design and construct electronic circuits for medical applications. After graduating I decided to study biology, focusing on cognitive sciences, tropical biology and social insects such as ants, three topics that fascinated me due to the emerging properties these complex systems exhibit. During my PhD I studied the potential environmental risks and public perception of genetically modified crops in centres of origin, places where humans first domesticated crops and where we still find an amazing diversity of wild crop relatives that are used for modern plant breeding. From traditional genetic engineering I turned my interest to safety issues in synthetic biology, which I see as the first attempt to really apply engineering principles to biology. In synthetic biology I could, for the first time, benefit from my combined background in engineering, biology and risk assessment. My work with synthetic biology and its biosafety ramifications brought protocells and xenobiology to my attention, two major research branches in synthetic biology that attempt to construct unfamiliar living systems, as shown in Table 1.

Synthetic Biology Subfield	Description	Degree of familiarity
DNA synthesis	Synthetic genes, artifical chromosomes, synthetic viruses, whole genome synthesis	↑
Bio-circuits	Genes and bioparts, biobricks, enhanced metabolic engineering, (e.g., artimisinin acid), evices, iGEM	
Minimal genomes	Top-down synthetic biology, reducing the genome of existing organisms, chassis for genetic circuits	
Protocells	Standard or alternative biochemistry, engineered phospholipids, cellular vesicles lacking key features of life, synthetic cells, bottom-up synthetic biology, manufacturing whole cells	
Xenobiology	Alternative biochemistry, Xeno nucleic acids (XNA), different bases, unnatural amino acids, changing the codon assignment of genes, development of novel polymerase and ribosomes, xeno-organisms, chemically modified organisms (CMOs)	↓

Table. 1: Different subfields and "unfamiliarity" in synthetic biology (Rasmussen et al., 2004; Benner and Sismour, 2005; O'Malley et al., 2008, Deplazes and Huppenbauer, 2009; Bedau and Parke, 2009; Schmidt et al., 2009).

While DNA synthesis is so far pretty much about reproducing existing genes and genomes (Gibson et al., 2010), engineered bio-circuits rearrange existing parts or genes (Ro et al., 2006; Canton et al., 2008), and minimal genomes are derived from existing simple organisms (Posfai et al., 2006). In other words, the first three subfields of synthetic biology are based largely on "life as we know it," and thus are "familiar" to us. The remaining two subfields, protocells and xenobiology, on the other hand, are attempts to show that lifelike properties do not have to be related to the biological systems that we find in nature, but could also

be realised in alternative ways. These proposed unfamiliar biological systems are either made from scratch (protocells) or using
entirely different bimolecular building blocks (e.g., as in xenobiology, using 6 instead of 4 bases in DNA, or nucleic acids where
desoxyribose or ribose are replaced by other molecules, producing so-called xeno nucleicacids) (Szostak et al., 2001; Benner and
Sismour, 2005; Marliere, 2009; Rasmussen et al., 2009; Schmidt,
2010). These unfamiliar biological systems can be seen as first examples of living technology; they are not only synthetic but also
different!

The final step to living technology came with the EC FP7
funded project TARPOL[1] where I looked into different synthetic
biology applications, including applications of microbial fuel cells.
These fuel cells are powered by organic materials and can be used
in so-called gastrobots, cyborg robots that use the electricity generated by the microbial fuel cell "stomachs." Other techno-bio
integrated systems that I found difficult to categorise (and which
therefore triggered my interest) were the remote controlled rats
and robots that are controlled by rat brain tissue (Talwar et al.,
2002; Warwick et al., 2010).[2] Even spiritual robots came to mind:
In 1999, two scientists, Ray Kurzweil and Hans Moravec, independently released books proclaiming that in the coming century,
our own computational technology will outstrip us intellectually
and spiritually (Kade, 2001). These systems led me to think about
expanding the principles of the living world to "platforms" other
than naturally evolved ones.

Also in my work on synthetic biology a bioethical question appears *vis á vis* the potential creation of synthetic life in a test
tube. Imagine we could construct bacteria from scratch, including
all cellular organelles and molecular components. Does that mean
we would encounter an ethical problem in doing so? To make
a long story short, in my mind we wouldn't. I have practically
no ethical reservations towards creating single-cell organisms in
the lab. On the other hand, however, imagine we could construct
a truly intelligent, emotional and self-conscious robot, that/who
would feel pain and pleasure. I would have much more ethical
doubts about creating such a self-aware robot than creating a living cell. Such a super-robot would exhibit some key features of

---

[1] See www.sb-tarpol.eu.

[2] A different approach to creating life out of plastic tubes is carried out by
Theo Jansen; see http://www.strandbeest.com/.

highly evolved organisms without being made out of traditional biological stuff. This *Gedankenexperiment* made it clear to me that living processes do not necessarily have to be based on "life as we know it." Once we become aware of the weak correlation between life as we know it and living processes as an abstract concept, a whole new cosmos opens up with amazing possibilities in implementing living processes in wet-, hard-, and software.

For some time I believed that all this was still too far away, more science fiction than a new science. However, in April 2009, after giving a one-hour tutorial at the 2nd European Science Foundation conference on synthetic biology, Mark Bedau approached me and invited me to become part of the ISSP Living Technology Working Group, which I gratefully accepted. The group consists of experts in robotics, biotechnology, IT, technology assessment and philosophy. Although I still think it is early, I very much appreciate the opportunity to brainstorm and discuss the advent of living technology in the ISSP working group with like-minded and visionary colleagues, not only because it is "nice" but also because I see living technology as an upcoming important science and technology area.

3. How is living technology related to overlapping or nearby research areas, such as nanotechnology, molecular biology, cloning and stem cell research, genetic engineering and synthetic biology? How is it related to social and technological systems such as social networks or information networks, such as the World Wide Web, cell phone networks and electronic banking networks?

In order to implement living technology two steps have to be taken: (1) achieving a thorough understanding of living processes, and (2) implementing these processes in technological systems of any possible kind, be it nanotech, IT, molecular biology or a combination of those.

To master step one we need all types of biology (molecular, micro-, systems, synthetic, etc.). The final benchmark proving that we really understand biological processes is to construct them, reflecting Richard Feynman's quote: "What I cannot construct, I do not understand." Construction of the first protocell, synthetic cell or xenobiological system that exhibits all characteristics of life will demonstrate our ability to understand and construct living systems. Once this has been achieved the next step will be the implementation of living processes in non-traditional biological matter, including, e.g., non-canonical biomolecules or even inorganic materials.

This two-step process will be accompanied by concomitant technological developments that improve our capabilities in engineering non-biological systems, such as intelligent autonomous robots, information networks, or manipulating matter at the nano-scale. Living technology will also benefit from a foreseen convergence of different technologies. These converging technologies (nanotechnology, biotechnology, information technology, cognitive science and robotics) will also feature a combination of wet-, soft-, and hardware; however, without the explicit goal of implementing living processes.

Living technology needs a large degree of interaction with other science and technology fields. Understanding living processes and applying them using engineering principles of converging technology will support the implementation of living technology.

4. What do you think are the most important open research questions about living technology, and how you think they should be pursued?

Living technology, in my mind, still has a long way to go before it will have a substantial impact on society. We still do not understand all the basic principles of life and their interconnectedness, much less how to engineer them. Many questions remain open, and in this early phase most of them have to be tackled by basic research. In parallel, I think it is also necessary to start thinking about the ramifications of future living technologies on society, the economy and the environment. I think the most important research questions and aims are:

- How do simple living organisms work?

- How can we reconstruct the characteristics of living systems in orthogonal biosystems with non-canonical biomodules (molecules), such as protocells or xeniobiological systems?

- Develop a programming language in order to deal with the abstract principles of life, to model and simulate living processes independently of a particular type of matter-structure.

- Develop a methodology to manage the convergence of different engineering disciplines, as a prerequisite to implementing the abstract rules of life in non-traditional systems (cells) involving wet-, hard-, and software.

- Engineer the characteristics of life into a converging wet-, hard-, and software system.

- Inquire about the societal, economical and environmental ramifications of living technologies throughout the different phases of development.

5. What do you consider to be the most interesting and important human or societal implications of research and development in living technology?

One the one hand there are the classical ethical, legal, and social issues (ELSI) that pop up in practically every new technology debate and will probably be discussed also in relation to living technology: safety, security, ethics, governance, and intellectual property rights (see, e.g., Schmidt et al., 2009). Pessimistic minds may argue like Bill Joy in his article "The future doesn't need us" (Joy, 2000) that some kind of self-replicating robot will take over the world. The typical nightmare scenarios include, for example, semi-living mobile intelligent robots that will patrol streets and various warfare theatres, making decisions themselves on how to interfere with prohibited human behaviour (USAF, 2009). A recent example of such autonomous systems is the Energetically Autonomous Tactical Robot (EATR), funded by DARPA. EATR is an autonomous robotic platform able to perform long-range, long-endurance missions without the need for manual or conventional re-fueling. EATR will be able to find, ingest and extract energy from biomass in the environment (e.g., battle fields). EATR triggered a debate that could give us a glimpse of how future living technology debates could roll out: following newspaper articles that EATR could feed on enemy corpses on battle fields, the producers of EATR issued a press release, saying: "In response to rumours circulating the internet on sites such as FoxNews.com, FastCompany.com and CNET News about a "flesh eating" robot project .... (we) would like to set the record straight: This robot is strictly vegetarian" (Robotic Technology, 2009). Also, I can easily imagine a large contribution towards human assistance, e.g., in a medical setting, after injuries or in case of old age. Geriatric intensive care or organ replacement will become easier to handle, opening also a market for human enhancement products and technical opportunities for various post- and trans-humanistic dreams (or nightmares, depending on your view).

Although many of these things are still science fiction, they might become reality in one or two decades, maybe even sooner. Think of a future living technology house that acts practically like a huge organism. It absorbs light on its roof and outer wall surface, converting it into energy (together with the insect- and rat-devouring electro-stomachs in the basement). Energy is stored in high-energy molecules in the walls and used up for self-repairing mechanisms (e.g., healing a crack in the wall), or maintaining optimal temperature for the human inhabitants, or lighting up a room at night with bio-iluminescence. The house could also engage us in a pleasant conversation, cheer us up and entertain us while it is whistling the kids to sleep next door. These and other features would make the house (and other living technology artefacts) "part of the family," in a similar way that a dog, cat or bird becomes part of the family.[3] As with pets we could become more attached to the wellbeing of the house and invest more time, money and energy into it, than we do with "traditional" friends, causing social interference. Living technology could make us dwellers in giant Tamagochis that keep us busy and docile, forgetting about the real world in the ultimate *panem et circenses*[4] scenario. Or it could enhance us, leading to more foresight and wiser people making better decisions for the benefit of humankind; it could also change the way we understand ourselves and nature.

Also, a number of philosophical challenges arise, especially regarding the transcendence of the characteristics of life to non-traditional or non-biological matter. In a way, should living technologies be successful, we will see a breakdown of the traditional boundaries between the living and nonliving worlds. Certainly it will rock our long held believe that life is something special or even magic, and that it is always part of nature (the naturally evolved type). Among the toughest question will be the ethical and legal

---

[3] To my mind pets can never be part of the family, only members of a group.

[4] I.e., "bread and games," which is a metaphor for a simplistic means of appeasement. The phrase is used to describe the creation of public approval or ignorance, not through excellent public policy, but through the mere satisfaction of the immediate, shallow requirements of a populace (e.g., entertainment, sports). It connotes the triviality and frivolity that defined the Roman Empire prior to its decline.

status of human-like intelligent robots, or non-human persons.

Let me end with a modified version of the so-called "trolley problem," a classical ethical dilemma:

"It is 2045, and a trolley steered by an autonomous living technology-enabled robot is running out of control down a track. In its path are five humans who have been tied to the track by a mad philosopher. Fortunately, there is a switch, which will lead the trolley down a different track to safety. Unfortunately, there is a single robot tied to that track. Will the robot flip the switch?"

About the Author: Dr. Schmidt is co-founder, board member and project leader at the Organisation for International Dialogue and Conflict Management (IDC) (www.idialog.eu) in Austria. He has an educational background in electronic engineering, biology and ecological risk assessment. His research interests include biosafety and technology assessment (TA) of novel biotechnologies including synthetic biology, management of genetic resources, public perception, science communication and film-making. Schmidt coordinated and participated in several international research projects (e.g., SYNBIOSAFE: Safety and Ethical aspects of Synthetic Biology: www.synbiosafe.eu), edited a book on synthetic biology, guest edited a special journal issue on societal aspects of synthetic biology, and produced two science documentary films for the general public. Before joining IDC Dr. Schmidt worked as researcher at the University of Vienna on several projects on risk assessment and risk perception, e.g., GM crops and nuclear fusion. He also worked as biosafety consultant for genetically modified crops in Southern Africa and China.

References

Bedau, M., & Parke, E. C. (2009). *The ethics of protocells: Moral and social implications of creating life in the laboratory.* Cambridge, MA: MIT Press.

Benner, S. A., & Sismour, A. M. (2005). Synthetic biology. *Nature Reviews Genetics*, 6, 533-543.

Canton, B., Labno, A., & Endy, D. (2008). Refinement and standardization of synthetic biological parts and devices. *Nature Biotechnology*, 26 (7), 787-793.

Deplazes, A., & Huppenbauer, M. (2009). Synthetic organisms and living machines: Positioning the products of synthetic biology at the borderline between living and non-living matter. *Systems And Synthetic Biology*, 3(1-4), 55-63.

Gibson, D. G., Glass, J. I., Lartigue, C., Noskov, V. N., Chuang, R. Y., Algire, M. A., Benders, G. A., Montague, M. G., Ma, L., Moodie, M. M., Merryman, C., Vashee, S., Krishnakumar, R., Assad-Garcia, N., Andrews-Pfannkoch, C., Denisova, E. A., Young, L,. Qi, Z. Q., Segall-Shapiro, T. H., Calvey, C. H., Parmar, P. P., Hutchison, C. A., Smith, H. O., & Venter, J. C. (2010). Creation of a bacterial cell controlled by a chemically synthesized genome. *Science*, 329(5987), 52-56.

Graham-Rowe, D. (2004). Self-sustaining killer robot creates a stink. *NewScientist*. Available online at http://www.newscientist.com/article/dn6366 (Accessed June 28th 2010).

Joy, B. (2000). Why the future doesn't need us. Wired Issue 8, 04. Available online at http://www.wired.com/wired/archive/8.04/joy.html (Accessed June 28th 2010).

Kade, R. (2001) Will Spiritual Robots Replace Humanity by 2100? Vol. 34, N. 1, pp. 82-83

Marliere, P. (2009). The farther, the safer: A manifesto for securely navigating synthetic species away from the old living world. *Systems And Synthetic Biology*, 3(1-4), 77-84.

O'Malley, M., Powell, A., Davies, J. F., & Calvert, J. (2008). Knowledge-making distinctions in synthetic biology. *BioEssays*, 30(1), 57.

Pósfai, G., Plunkett, G., Fehér, T., Frisch, D., Keil, G. M., Umenhoffer, K., Kolisnychenko, V., Stahl, B., Sharma, S. S., de Arruda, M., Burland, V., Harcum, S. W., & Blattner, F. R. (2006). Emergent properties of reduced-genome escherichia coli. *Science*, 312 (5776), 1044-1046.

Rasmussen, S., Bedau, M. A., Chen, L., Deamer, D., Krakauer, D. C., Packard , N. H., & Stadler, P. F. (2009). *Protocells, bridging nonliving and living matter*. Cambridge: The MIT Press.

Rasmussen, S., Chen, L., Deamer, D., Krakauer, D., Packard, N., Stadler, P., & Bedau, M. A. (2004). Transitions from nonliving to living matter. *Science*, 303, 963-965.

Ro D. K., Paradise, E. M., Ouellet, M., Fisher, K. J., Newman, K. L., Ndungu, J. M., Ho, K. A., Eachus, R. A., Ham, T. S.,

Kirby, J., Chang, M. C., Withers, S. T., Shiba, Y., Sarpong, R., & Keasling, J. D. (2006). Production of the antimalarial drug precursor artemisinic acid in engineered yeast. *Nature*, 440, 940-943.

Robotic Technology Inc. (2009). Energetically autonomous tactical robot (EATR) project. Available online at www.robotictechnologyinc.com/index.php/EATR (Accessed June 28th 2010)

Schmidt, M. (2010). Xenobiology: A new form of life as the ultimate biosafety tool: *BioEssays* 32(4), 322-331.

Schmidt, M., Ganguli-Mitra, A., Torgersen, H., Kelle, A., Deplazes, A., & Biller-Andornoet, N. (2009). A priority paper for the societal and ethical aspects of synthetic biology. *Systems And Synthetic Biology*, 3(1-4), 3-7.

Szostak, J. W., Bartel, D. P., & Luisi, P. L. (2001). Synthesizing life. *Nature*, 409, 387-390.

Talwar, S. K., Xu, S., Hawley, E. S., Weiss, S. A., Moxon, K. A., & Chapin, J. K. (2002). Behavioural neuroscience: Rat navigation guided by remote control. *Nature*, 417, 37-38.

USAF (US Air Force) (2009). Unmanned Aircraft Systems Flight Plan 2009-2047. Available online at http://handle.dtic.mil/100.2/ADA505168 (Accessed June 28th 2010).

Warwick, K., Xydas, D., Nasuto, S. J., Becerra, V. M., Hammond, M. W., Downes, J. H., Marshall, S., & Whalley, B. J. (2010) Controlling a mobile robot with a biological brain (review paper). *Defence Science Journal*, 60 (1).

16

Kasper Stoy

Associate Professor

Maersk McKinney Moller Institute, University of Southern Denmark

---

1. In what sense do you find it meaningful to talk about "living technology?"

Living technology provides us with an overarching vision of technology that we can strive towards in our individual areas of research: a technology that is not dead and detached from the natural world, but due to its lifelike properties is intimately integrated into it. Living technology ultimately may break down the barrier between the artificial and the natural world.

In addition to being a vision, living technology also facilitates interaction between researchers from different fields that typically have no reason to interact because their fields have completely different challenges and problems. This interaction is important because it may make us see our work as part of a greater whole and realize potential synergies between the emerging fields of living technology.

For me this general motivation was enough to engage with the living technology community, but for the concept to have a lasting interest it is important that it also lead to scientific break-through. Therefore, the real test of the concept for me is if related fields to modular robotics, which is my field, can help us address some of our fundamental challenges. Modular robotics (and robotics in general) is a science of integration, and its progress rests on the development of new components and of course on our ability as robot engineers to synthesize them into complete robots. Examples of breakthroughs that may come out of living technology are components that are more resilient to natural environments and can be powered and self-replicated from resources available in the environment. This may in turn allow us to synthesize robots that are more autonomous and thereby more useful. If, by participating in the living technology community, I can motivate the development of such technology for robotics, living technology is a useful concept for me; if not, living technology as a concept may be stranded because it cannot merge disciplines and fields, because the underlying research questions and challenges are completely different.

2. How does your research relate to living technology, and why were you initially drawn to do this work?

My background is in modular robotics, which is a niche in the larger research areas of robotics and embodied artificial intelligence. In robotics and artificial intelligence there has always been a focus on mimicking aspects of living systems. This goes all the way back to Karel Capek, who in 1920 coined the word 'robot', which comes from the Czech word 'robota' which means something like "serf." In Capek's world a robot was a human without a soul. A human stripped of everything not needed to perform work. The word 'robot' was applied later to machines, and the robots of today typically do work as "slaves," performing dull, dirty, or dangerous jobs. However, one aspect of living systems that has not been captured so well in robots is their ability to deal with complex, dynamic environments. Robots, in contrast, typically require structured, static environments in order to function. It has therefore been natural for robotics researchers to look to living systems for inspiration on how to create robots that are able to function in more natural environments. In living technology we aim to take this tradition of biological inspiration one step further, and go from robots that mimic living systems to systems that actually live. By living I mean that they are able to use available resources in the environment for energy production and self-replication. Living robots potentially will be a significant improvement over existing robots due to their ability to maintain themselves.

I feel that modular robotics, my own research area, represents a natural path from robots that mimic life to robots that live. The idea behind modular robotics is to transfer the concept of multicellular organisms to robotics. That is, instead of building robots as monolithic pieces of mechatronics we build robots from robotic cells, or modules as we call them. This modular approach potentially provides us with the advantages of versatility, robustness, and reduced cost. Versatility comes from the fact that modules can be assembled in many different ways, and thus a wide variety of robots can be built from the same basic set of modules. Robustness arises because there is no single point of failure: If a module fails the remaining modules can take over its functionality. Finally, while the basic modules are relatively complex, they can be mass-produced and thus made cheap relative to their complexity.

In the context of living technology, modular robotics may become useful because in the field we have established how to combine simple elements into complete robots. This becomes relevant if other areas of living technology can provide the simple elements, e.g., maybe in the form of protocells. If this is the case, we have a path that leads straight from molecular chemistry to robotics, which would have significant consequences in terms of these systems' ability to live. The chance to explore the feasibility of this has drawn me to the living technology community. It is also always inspiring for me to meet people who think differently, and it was therefore easy for me to decide to sign up for my first workshop on living technology.

3. How is living technology related to overlapping or nearby research areas, such as nanotechnology, molecular biology, cloning and stem cell research, genetic engineering and synthetic biology? How is it related to social and technological systems such as social networks or information networks, such as the World Wide Web, cell phone networks and electronic banking networks?

Living technology is basically a unifying vision for all these more or less related areas and may bias what the interesting questions are in each research area. The concept of living technology also highlights potential links between these research areas that may be worth exploring. I think maybe this is really what sets the concept of living technology apart, that it is able to unify otherwise unrelated areas.

However, I think the research areas of living technology fall in two categories that do not necessarily have much in common, and I think that a marriage of the two will be less fruitful. Here I think about the physics-based areas such as nanotechnology, modular biology, and so on, as opposed to the information-based technologies such as the World Wide Web, cell phone networks, etc. From the perspective of artificial life it is natural to combine them, because there has been a tradition of treating hard, wet, and soft artificial life equally. However, as a roboticist that has followed the embodiment movement across many disciplines, from cognitive science to artificial intelligence and robotics, it is less clear why they should be combined. An important understanding that has emerged over the last decade is that the physical implementation matters, and indeed some will go as far as to say that the lack of progress in artificial intelligence and robotics is due to negligence of the need for intelligent systems to be physically embodied. So where I see a link between hard and wet artificial life, I think the link to soft artificial life is less interesting.

4. What do you think are the most important open research questions about living technology, and how you think they should be pursued?

An open question is whether living technology is actually a useful concept. Will the conversations between the fields related to living technology be of benefit to the individual research areas, and more ambitiously will these areas be able to come together and produce technology that is truly living?

From a robotics perspective the most important question is if some of the more fundamental areas of living technology can provide or be developed to provide pieces with enough functionality to enable us to move robotics forward. Concretely, can living technology provide a new generation of robotics components that are more appropriate for robots working in natural environment? E.g., can it provide components that are resilient to dirt, water, changing heat and light levels, etc., and are robust enough to handle repeated collisions (intended and non-intended) with the environment? Can it provide components that can be self-replicated from and powered by resources available in the environment? Finally, can these new living technologies be synthesized into living robots that can perform meaningful tasks in complex, natural environments while not requiring the attention of a human for refuelling, repair and production?

It is not clear how to answer these questions, but in the short term it is important to disseminate the knowledge of the sub-fields of living technology to the broader audience of living technology, to see if answers can be found or developed.

5. What do you consider to be the most interesting and important human or societal implications of research and development in living technology?

Living technology, if realized, will pose both new opportunities and dangers to human society. However, to predict which implication is most important, given that living technology is a vast collection of related, powerful technologies, is impossible. In robotics the most powerful technology would be robots that actually can take care of themselves while performing their tasks. That is, robots that are energetically autonomous and can perform self-repair or even self-replicate from available resources in the environment. It is clear that if realized such robot technology would have implications for humans as individuals and society in general, but it is beyond me to predict or even speculate about precise implications because it still remains in the realm of science fiction.

Given this position I also find it questionable to approach the public using the concept of living technology. When not even we as researchers can predict a likely outcome, how is the general public supposed to? The discussion with the public will then rest purely on fears and imagination, and not on scientific fact, which may sidetrack the discussion of these potential useful technologies as happened with nuclear power, genetically modified crops, and so on. This is something we need to avoid, and I therefore feel living technology as a concept is not so useful for engaging the public. Rather, when it is realized, positive usage of a specific living technology will serve us better because we can then present a positive view of living technology together with a fair warning. A cornerstone of this will be to provide knowledge to the public about how the specific piece of living technology works and what its limitations are, so that discussions do not regress into arguments based on fear rather than reality.

About the Author: Kasper Stoy is an associate professor at the Maersk McKinney Moller Institute, University of Southern Denmark (USD), and a co-director of USD's Modular Robotics Lab. He received his MSc in computer science and physics from the University of Aarhus, Denmark and his PhD in computer systems engineering from USD in 2003. He spent a year of his PhD studies at USC's Information Sciences Institute, CA, USA and has been a visiting scholar at Harvard University. Prof. Stoy is the author of the book *Self-Reconfigurable Robots: An Introduction* (MIT Press, 2010), and has published more than forty papers of which three first-author papers received awards. He organizes international workshops on modular robots, serves as reviewer for IEEE conferences and journals, including the *International Journal of Advanced Robotics, Journal of Autonomous Robots, Journal of Simulation of Adaptive Behaviour* and several more. He also developed the first version of the Player component of the multi-robot simulation tool Player/Stage, which is the most widely used simulation in this field worldwide, and co-founded the company Universal Robots. He currently manages the "Morphing Production Lines" research project funded by the Danish Research Council for Technology and Production and is USD's PI on the EU project Locomorph.

17

Giuseppe Testa

Principal Investigator

Laboratory of Stem Cell Epigenetics, European Institute of Oncology

---

1. In what sense do you find it meaningful to talk about "living technology?"

The proposal to group under the heading of "living technology" all technologies that incorporate the features of life in their operations is meaningful first of all insofar as it is an attempt to transgress disciplinary boundaries. The current distinction between technologies that manipulate living organisms directly (broadly referred to as biotechnologies) and those that engineer nonliving matter is a legacy of the historical trajectories that moulded the developments of the various technoscientific disciplines of our age. Life as a distinct phenomenon, as something that was worth studying precisely because it had unique properties, has spurred human curiosity since the very beginning, as far as we can tell. And even the frontiers of the molecular life sciences today can be said to be descendants of the same intellectual pursuit that we trace back to the earliest civilizations of which we have a record. From Aristotle's insights to the ambitions of synthetic biology much has changed of course. The science of evolution has given us a broad explanatory framework to understand how life changes, enabling us to interrogate it retrospectively and, to an increasing extent, affect it prospectively. And molecular biology has given our gaze an unprecedented degree of depth as well as breadth, offering us mechanistic insight into at least some of the levels of regulation that underlie living phenomena. But through all these epistemic shifts and technological revolutions, the intellectual pursuit has stayed by and large the same: the study of life itself.

What living technology now attempts to do is expand this domain of inquiry in a radical manner. We could say that instead of the study *of* life itself, "living technology" foresees and advocates

study *with* life and *from* life itself. Rather than "simply" changing
life from within (the thrust of molecular life sciences), it proposes
to enlarge the very scope of life, by infusing a variety of technological systems *with* features and principles that are imitated,
borrowed or extracted *from* life itself. Here lies its *telos* and the
core of its intellectual challenge: Now that we have understood
what sets life apart at least to a certain degree of molecular detail, we can start to bestow living properties upon what surrounds
us that is not living. It is a Golemic project, in the fullest sense
of the word; the injection of plastic and self-assembling resilience
into the stiff endurance of the inert world.

2. How does your research relate to living technology, and why were you initially drawn to do this work?

My lab at the European Institute of Oncology studies the epigenetic mechanisms that enable lineage commitment and their
aberrations in cancer. Our aim is to understand how genomic
programs are progressively deployed in order to enable cell fate
transitions through both physiologic and aberrant development
and what are the chromatin regulatory mechanisms that coordi-
nate their deployment. In particular, we focus on the methylation
of histone H3 on lysine tails 4 and 27, respectively mediated by
the Trithorax (Trx) and Polycomb (PcG) protein families, since
established genetic evidence and more recent molecular studies
indicate that these two chromatin axes are central to the programming of genomes that underlie the establishment and maintenance of differentiated cell states. Not surprisingly, aberrations
in these pathways have also emerged as important determinants
or modulators of tumors, hinting at common regulatory circuits
that preside over stem cell physiology and that are perturbed or
hijacked in oncogenesis. Finally, changes in these posttranslational
modifications are also prominent in the epigenetic rewiring that
recently reversed Waddington's unidirectional slopes, namely the
reacquisition of pluripotency from differentiated cells through nuclear transfer or the expression of few pluripotency factors. Hence,
work in my lab is articulated in three complementary lines of research that investigate PcG and Trx function in: Ii) the physiology
of genome programming during differentiation using neurogenesis
as a model system; (ii) the aberrant genome programming that
accompanies tumorigenesis; and (iii) the controlled genome reprogramming that mediates induced pluripotency.

As such, this work is highly relevant to one subset of living technology, namely the engineering of defined cell types, which requires an understanding of their physiological development and the ability to rewire the transcriptional and post-transcriptional networks in order to reassign cell fate. Over the last two decades, evidence from various experimental systems has shown that cell fate can be changed through the forced expression of transcription factors (TF) (reviewed in Graf and Enver, 2009). Unexpectedly, few TF suffice to affect radical cell fate reassignment, as exemplified in the pathbreaking generation of induced pluripotent stem cells (iPSC) from adult somatic cells (Takahashi and Yamanaka, 2006). Collectively, these outcomes provide proof of principle that we can generate at-will defined cell types from any individual. I would argue that this represents a foundational resource for that strand of living technologies that aims at restoring or enhancing bodily functions, in at least two ways. First, the availability of virtually endless supplies of defined cell types, either normal or sourced through cell reprogramming from individuals affected with chronic diseases, yields an unprecedented opportunity to make human physiology and physiopathology amenable to high-throughput genetic and chemical screens. This in turn should pave the way to a host of molecular interventions that alter living processes from within the body in more precise and possibly radical ways. Second, the reprogrammed cells could themselves be the seeding living elements in more ambitious applications of "living technology," such as the generation of engineered tissues or organs that integrate nanotechnological devices or electronic circuits into new functional assemblies.

There is also a second reason why I consider my work highly relevant to living technology and why, in turn, I find the engagement with this set of ideas deeply rewarding. I have developed a parallel career in Bioethics and Science and Technology Studies (STS), out of a long-standing personal interest as well as out of the conviction that practicing life scientists, especially those working at the forefront of genetics and mammalian developmental biology, need to engage thoroughly with the ethical and social implications of their work. This interest began during my PhD at EMBL, where I contributed to establish the Forum on Science and Society, and grew into a true scholarly pursuit when I was awarded in 2003 the Branco Weiss Fellowship Society-in-Science, a pioneering initiative based at the ETH Zürich and aimed at fostering interdisciplinary research by practicing life scientists on the social dimensions of current life science. I took this fellowship as a unique opportunity to pursue in parallel my scientific project at the bench and my scholarly inquiry into the social framing and uptake of epigenetic research. Thus, while firmly anchored on my postdoc project on the genome engineering of embryonic stem (ES) cells, I pursued in parallel a Master in Bioethics and Biolaw at the University of Manchester by distance learning and became a visiting fellow in the Program on Science, Technology and Society at the Harvard Kennedy School of Government. This expertise ushered in a great opportunity to co-found, together with Epistemologist and Bioethicist Giovanni Boniolo, a new PhD program called "Foundations of the Life Sciences and their Ethical Consequences" at the European School of Molecular Medicine and the University of Milan. This PhD is the first example in Europe of an interdisciplinary PhD, aimed at scholars from both the humanities and the life sciences, that is fully integrated within the research premises of a life sciences institute. The aim is to foster a new range of professionals equipped with the cross-disciplinary skills that the prominence of biology in the public sphere calls for. And for the reasons I describe below, I believe that a dual competence in the molecular life sciences and the humanistic subjects that interact with them can be a useful example to orient the training of the future living technology scholars in a way that takes on board, from the inception, the mutual development of the technology and the social order within which it unfolds.

3. How is living technology related to overlapping or nearby research areas, such as nanotechnology, molecular biology, cloning and stem cell research, genetic engineering and synthetic biology? How is it related to social and technological systems such as social networks or information networks, such as the World Wide Web, cell phone networks and electronic banking networks?

By its very definition, living technology includes, and in fact critically builds upon, developments in molecular biology and genetic engineering. As briefly anticipated above, the aim of living technology is to endow various technological systems, from robots to softwares to Internet networks, with distinctive features that are sourced from or "simply" inspired by or copied from living organisms. Not all applications will require the actual transfer of organisms, molecules or molecular systems (defined here as sets of molecules that operate in a predefined arrangement) into technological systems that are currently outside of the life domain. In fact, the ultimate challenge of living technology, its intellectual but also very practical bet, so to speak, is that we will be able to model features of the living without resorting to components that were sourced from living entities. And the definition of primary living technology captures precisely this more radical instantiation of the endeavor (Bedau et al., 2010). Yet, it is clear that although many applications will, by ambition or necessity, not rely on the building blocks of life, the very idea of modelling living features requires that we acquire a full grasp of at least the most relevant mechanisms that underlie those features. That is why I believe that molecular biology, even when it will not provide a repository of building blocks, will still be central to the whole pursuit by providing models and conceptual resources.

However this point brings us also to an interesting epistemic tension, which I believe is at the core of living technology. It is a tension that originates in molecular biology as such, but that will reverberate to this emerging set of interconnected disciplines. It is the tension between the accumulation of mechanistic details on individual aspects of a cell's or tissue's function, which is still the dominant activity, and the attempt to capture directly the overall features of those functions by looking at the system level. The thrust of systems biology is precisely the ambition to understand cells and eventually organisms by analysing large data sets in which, almost by definition, the details are lost or made irrelevant. They are not irrelevant in the sense that they do not contribute to the phenotype under investigation; they very much do, otherwise there would be no point in accumulating them. But the point is to capture the so-called emerging properties of the system as such, which result from but cannot be reduced to the individual components even when described in full detail (though this is admittedly the still-unresolved sticking point that spurs the tension between reductionistic and holistic approaches). It is a juxtaposition of two modes of understanding, of two ways of assessing explanatory power. If systems biology delivers what it is poised to, then it may very well be that the pace of advance in living technology, at least in the primary living technologies, will be faster than anticipated. This is because we will not need a full understanding of all or even most components of a living process in order to model it, if we have understood sufficient aspects of it that can be translated into other material contexts.

4. What do you think are the most important open research questions about living technology, and how you

think they should be pursued?

There are many research questions that are still very open, but I believe that the most pressing one is conceptual in nature, and concerns the very definition of this emerging discipline. The question is to what extent the analogies between the different systems, from self-repairing materials to protocells to adapting softwares to cities and social networks, reach beyond the level of useful metaphors and become instead powerful, perhaps even necessary tools in order to let "living technology" mature into a coherent though multilayered set of disciplines.

In terms of specific research fields, one outstanding question is the extent to which living organisms as we currently know them will actually integrate, in flesh and blood so to speak, the new artefacts generated through living technologies. I am thinking especially about a variety of adapting and self-repairing biomaterials, which could have a significant impact on medicine if they can be integrated in meaningful ways into the human body. But I am also thinking about other examples from the secondary living technologies, in particular engineered cells that could be used to restore compromised organ functions or to enhance physiological levels of function, for example through the adaptive delivery of defined chemicals of through the adaptive interaction with aging tissues. For all of this to occur, a substantial amount of research should be devoted to the *in vivo* modelling of these interactions, and a high priority is the generation of the relevant models and the definition of a roadmap with the milestones that are considered necessary in order to approach this goal.

In other words, whereas there is no doubt that the current framing of living technologies entails, by definition, a close and multifaceted interaction between living organisms and the "new" forms that embody the features of life into various material or less material substrates, it remains very much open the extent to which these interactions will also unfold within the very same organism, leading to a factual coalescence of living organisms and the living technologies they have produced. We should keep in mind that for some application this point may become irrelevant. This is the case, for example, of those interfaces between humans and robots or human and softwares, in which, as long as this interaction enhances or repairs or expands human capacities, it is the end result that will matter. The fact that it will have been reached without an actual physical connection between the human subject and its surrounding technologies will be a minor point.

In other contexts, however, the possibility of integrating into living tissues (and I mean specifically human tissues for the sake of this discussion) will be crucial to reach the desired effect, and it is this type of living technology which, in my opinion, will necessitate most a long-term planning of resources and models in order to bring it to fruition.

5. What do you consider to be the most interesting and important human or societal implications of research and development in living technology?

Given what I have indicated above as the Golemic scope and depth of living technology, it is clear that its development will raise several ethical, social and legal issues, and it is important that the proponents of this scientific and technological shift are including these concerns from the outset as part and parcel of the roadmap that should facilitate this technological transition.

The first point I would like to raise concerns the timescale of these developments. I think that in the next 5-10 years we will not confront major, paradigm-shifting or potentially unsettling developments. To be sure, there will be tremendous progress in the various disciplines that converge into living technology, if we extrapolate from the pace of development in the last three decades. Yet I do not believe that, at least in the biomedical area, the seamless integration of living organisms with or within technological systems that are currently nonliving will have reached the depth that constitutes the very aim of living technology. I base this timescale skepsis on the empirical experience from the last twenty years in two closely related areas of biology, molecular genetics and cell biology. Both have progressed tremendously. We can now sequence entire genomes in a tiny fraction of the time and at a ridiculous fraction of the cost that was initially needed to produce the first draft of the human genome in the year 2000. In the course of this massive upscaling, we also have learned, however, that genome regulation is orders of magnitude more complex than our previous models predicted. The more we gaze into the noncoding potential of our genome, the more we discover mechanisms and options for regulation that simply had evaded our previous system of thought. And it is clear that it will take many years to grasp the full functional implications of these elements in the various cells that make up complex organisms. A similar story can be told for the unprecedented developments in the field of molecular cell biology, where we can now turn virtually every somatic cell into a pluripotent stem cell or even directly into a different kind of somatic cell. Yet, in so doing, we are also beginning to appreciate the subtleties of regulation that underlie these process, and with them also our current limits in imitating the physiologic process of genome programming that underlies the acquisition and maintenance of cell fate. None of this should inspire either pessimism or disillusion. But it should invite a healthy dose of humbleness and a realistic appreciation of the challenges ahead.

Given the scope and format of this book, it does not strike me as particularly useful, at this stage anyway, to delve into the individual ethical and social problems that will accompany the development of living technology. I would therefore like to concentrate on what I consider the central theme that should guide our thinking across the board of all potential ethical and social problems. This theme is public participation. I am not speaking here of various efforts at informing the purportedly lay public about living technology. While it is obvious that all – or at least most of us – would like to be better informed about many things, a state of enduring and inevitable ignorance has become an ever more salient trait of the human condition in the age of pervasive technology. Countless empirical works from the field of Science and Technology Studies (STS) have shown unequivocally that, when it comes to reception of technology on the side of the public, knowledge and information lead certainly to more constructive and articulated debates but not necessarily to larger degrees of acceptance. But there is a more important normative point as well. If we are serious about our democratic commitment, if we are serious about endorsing "rule by the people" in a time in which technology is integral to virtually any aspect of our daily lives, we should not simply tolerate, but in fact vigorously promote the engagement of citizens into the decision processes that guide technoscientific development. Thus, the scholarly pursuit, as well as the normative appeal, to move from public understanding *of* science to public engagement *with* science, from acceptance or rejection downhill to active participation uphill, should become integral to the development of living technology. In fact, I'd like to propose that this key transition from understanding to engagement find precisely in living technology a great opportunity to unfold and mature, for two reasons. First, as the collective work of this volume shows, living technology is very much in the making, the far-reaching connections between its various constituent subdisciplines still to be implemented. Hence, this is the right time to engage with the important questions on how this technological transition should occur, in terms of scientific, political and economic directions. For example, the very aim of bestowing living properties upon nonliving technological systems gives ample reason to reflect. On the one hand this goal is eminently attractive, because features like self-sustainment, self-repair and self-propagation would make many technological systems immensely more efficient and at least prima facie much better at interacting with existing living systems. On the other hand, self-sustainment, self-repair and self-propagation also confer on living organisms their various degrees of autonomy. How to go about this thin trail between infusing technology with enough life so as to enhance it while preventing it from acquiring too much control of its own? These are watershed questions, and bring us to the second reason for engaging with the broad implications of this technology now and here. Living technology is in fact by definition pervasive, in that its aim is to create a virtually all-encompassing system of interconnected technologies based on life. And precisely because of its depth and breadth of impact, citizens should be empowered to co-shape its course along with the many scientists and scholars who will advance its progress. Needless to say there is no easy, off-the-shelf solution, and certainly not a solution that will fit magically every geographical area, despite the mantra of globalization. But as I have argued recently, we need to rethink, on the basis of the available and already rich experience, the institutions that mediate citizens' participation, in order to promote the emergence of living technologies that are both scientifically and socially robust (Nowotny and Testa, 2009). In fact, I believe that the social experimentation of devising new modalities and new institutions of public participation should be seen as an integral element in the conflationary vision of living technologies.

About the Author: Giuseppe Testa holds an MD from the University of Perugia Medical School, a PhD from the European Molecular Biology Laboratory in Heidelberg, and an MA in Health Care Ethics and Law from the University of Manchester. He heads the Laboratory of Stem Cell Epigenetics at the European Institute of Oncology in Milan, focusing on the role of histone modifying enzymes in stem cell differentiation, both in physiological development and in cancer. An awardee of the Branco Weiss Fellowship "Society in Science," he was a visiting fellow in the Program on Science, Technology and Society at the Harvard Kennedy School of Government and at the Berlin Institute for Advanced Studies. In 2006 he co-founded in Milan with Giovanni Boniolo the interdisciplinary PhD program of the European School of Molecular Medicine on Life Sciences, Epistemology, Bioethics and Society. His bioethics and STS scholarship focuses on the co-production of body lineages in the public sphere and the emergence of new participatory trends in biomedicine. He was awarded the Roche Prize for leading bioscientists of the next decade and serves on the Ethics and Public Policy committee of the International Society for Stem Cell Research. Both his scientific and bioethics/STS work has been published in leading peer-reviewed journals, including *Nature, Nature Biotechnology, Science, Cell, Cell Stem Cell, PLoS One, Bioethics* and *Science as Culture.*

References

Bedau, M., McCaskill, J. S., Packard, N. H., & Rasmussen, S. (2010). Living technology: Exploiting Life's principles in technology. *Artificial life,* 16, 89-97.

Graf, T. & Enver, T. (2009). Forcing cells to change lineages. *Nature,* 462(7273), 587-94.

Nowothy, H. & Testa, G. (2009). Die gläsernen Gene: Die Erfindung des Individuums im molekularen Zeitalter Berlin: Suhrkamp Verlag (Naked Genes: Reinventing the Human in the Molecular Age). Forthcoming in 2010. Cambridge, USA: MIT Press.

Takahashi, K. & Yamanaka, S. (2006). Induction of pluripotent stem cells from mouse embryonic and adult fibroblast cultures by defined factors. *Cell,* 126(4), 663-76.

18

Mihaela Ulieru

Professor

Adaptive Risk Management Laboratory, University of New Brunswick

---

1. In what sense do you find it meaningful to talk about "living technology?"

There are certain unique properties of living systems that we are seeking to imitate in the quest to design systems and organizations with high agility, which can dynamically change, exhibit emergence, and self-organize to adapt, reconfigure and quickly respond to unexpected societal and environmental demands. Applications for such technology abound in a large variety of problem domains in our ever-changing world, since several critical problems are due to the rigid structure of our social, political and economic systems. These systems do not allow for adaptation and agility of response to unexpected, emerging needs, but rather act as roadblocks on the path to implementing effective necessary solutions. In our quest to make our world more sustainable and resilient, we need to acknowledge the limitations of our legacy of (institutional and critical) infrastructures, which were built in the industrial age. A deep understanding of the intimate mechanisms behind the properties that make a living technology "alive" will point to the need to change the cultures in our organizations and revolutionize how we live and work through a radical shift in the way we interact with (and within) our socio-politico-economic systems, as well as with the natural environment: from being constrained and dependent on technology and institutional structures to having them fuel our creativity and innovative potential in a proactive, anticipatory manner.

Several emerging technologies related to the pervasiveness of information and communication technologies (ICT), combined with advances in nano- and biotechnology, enable spontaneous linkages between people, systems, infrastructures and "things" to create "living" ecosystems that can configure and reconfigure as needed. Such technology can offer process and production continuity in all areas and aspects of life and work. In an ever-growing system of systems linked by intelligent communication networks, *softbots* (autonomous software robots that emulate human abilities, such as making decisions, discovering knowledge and performing tasks on behalf of the user in cyberspace) and *nanobots* (autonomous nanomachines exhibiting human-like behaviour to perform tasks in environments inaccessible to the human, such as carrying drugs to various parts of the body through the blood vessels) start interacting to create "societies in Cyberspace." These Cyber-Physical Ecosystems exhibit a life of their own in an emerging parallel universe, bridging the physical and the virtual, merging us with our "things," transforming the way we live and work, and augmenting our abilities in unprecedented ways. Examples range from self-reconfiguring manufacturing plants and self-stabilising energy grids to self-deploying emergency taskforces, all relying on myriad mobile devices, software agents and human users that would build an ecosystem bringing the right skill, tool, or competence at the right time for the right task on the sole basis of local rules and peer-to-peer communication. In such opportunistic ecosystems, distributed systems at various levels of resolution, ranging from single devices to spaces, "living" in an "Internet of Things" connected via communication networks and enterprises, are brought together into a larger and more *complex* "system of systems," in which the individual properties or attributes of single systems are dynamically combined to achieve an emergent desired behaviour of the synergetic ecosystem.

The dramatic progress of Cyber-Physical Ecosystem technologies is envisioned to reach unanticipated levels of complexity, beyond the boundaries of the disciplines that conceived their components. This challenges the traditional engineering school of thought in disruptive ways, given that, by their very nature, these technological ecosystems cannot be defined a priori, but rather *emerge* from the interactions between individual systems (and people), interactions facilitated by the communication networks. The unanticipated levels of complexity exhibited by these ecosystems are transcending the boundaries of the sciences under which these artefacts were conceived and incrementally crafted. This requires drastic revision of the traditional top-down perspective on system design and control, which aimed at imposing order *exogenously*, telling each element of the system what to do at every step through predetermined strategies, and assuming that all possible situations the system might confront are knowable in advance. Living technologies offer inspiration from natural systems for a new way of designing and engineering such digital ecosystems that are being interwoven into our world's infrastructures, from governance structures to critical infrastructures, endowing them with an intrinsic ability to self-replicate, evolve and adapt to support us, respond to the demands of an ever-changing, unsafe and convoluted world.

2. How does your research relate to living technology, and why were you initially drawn to do this work?

During the early 1990s the Japanese government indentified a clear need to create intelligent manufacturing systems endowed with agility and resilience able to quickly reconfigure in response to unexpected customer production demands. To implement such systems, the Holonic Manufacturing Systems (HMS) consortium took inspiration from Arthur Koestler's seminal book *The Ghost in the Machine*, which introduced a generic paradigm for designing self-organizing systems at all scales by emulating general principles of how the universe self-organized in a manner that enabled the emergence of life. The ultimate goal was to discover the laws that can create the premises for "living" manufacturing systems endowed with self-* abilities (which denotes a series of self-reflexive phenomena: e.g., self-replicaiton, self-movement, self-healing, self-organizing, self-repair, etc.), that can self-repair and self-configure around dynamically changing production needs. In conjunction with this project, the HMS consortium was concerned with the design and development of intelligent robotic systems that emulated human abilities and could create autonomous "societies" "living" on the manufacturing shop floor, which could replicate and repair themselves and each other. High-level coordination was necessary through production planning within the manufacturing enterprise, and in turn production depended on the supply chains managed with outside partners, which also had to be implemented in an ecosystem-like manner. The principles of living systems were applied at all scales, to create the premises for the manufacturing technologies themselves to become "living production ecosystems." Specifically, production processes are information-rich, and the dynamics of the information infrastructure is the tool for carrying it out both at individual locations and across the global environment. The electronic linking implies that work matter (or critical parts of it) is being transferred across virtual locations via the dynamic service environment, which supports organizational information that in turn can mirror social organization.

For the implementation of this technology, the latest advances in distributed artificial intelligence were used, namely software agents coupled with the latest ICT advances, which resulted in the creation of a dynamic service environment of intelligent mobile agents regulating production in a distributed manner by mirroring how the nervous system regulates a living organism. As member of the HMS consortium I collaborated with the Foundation for Intelligent Physical Agents (now part of the IEEE standardisation efforts of the IEEE Computer Society) to design this dynamic service environment as a standard for the implementation and deployment of distributed autonomic manufacturing systems. The mobile agents that were carrying the command and control function across the manufacturing plant, being gradually equipped with more and more intelligence and human-like abilities, began having a life of their own in cyberspace, thus creating "living" societies from which the concept of an "eSociety" resulted. The result of my five-year quest as the Canada Research Chair in eSociety (2005-2010) was a generic methodology for designing self-*-distributed systems for coordinating various activities, ranging from manufacturing production and energy distribution to online community building and virtual organizations that dynamically link the right skill at the right time at the right place for the coordination of streamlined and agile emergency response and business operations.

3. How is living technology related to overlapping or nearby research areas, such as nanotechnology, molecular biology, cloning and stem cell research, genetic engineering and synthetic biology? How is it related to social and technological systems such as social networks or information networks, such as the World Wide Web, cell phone networks and electronic banking networks?

Implications of such broad-scale, networked, independent yet collaborative agents for societal functions such as trade, commerce, military applications, energy, transportation, health, education and entertainment are arguably extremely significant nowadays. Such progress in networked systems has transformed our world to the extent that it can be argued that very large numbers of entities (mostly modelled as software agents) now exist virtually in a universe of networked information, with distinct parallels to the universe we can normally apprehend through natural and extended human senses. Implementing an organizational structure (modelled after a real-life, complex adaptive system) into software using the multi-agent systems framework opens the perspective of regarding the Internet as the equivalent of multiple societies of agents comprising a virtual (digital) ecosystem that emulates different contexts of the real world, cloned in software by abstracting various functions and abilities of real entities as needed to fulfil the specific contextual purposes addressed by the ecosystem..

Such digital ecosystems are built on advances in: sensor, wireless and optical networks; secure software design; models for social networks and economic behaviour; resource management and cloud computing; and complex systems theory to address fundamental challenges in the design and operation of new, massive-scale, complex computing and communications systems and apply these new systems to enable the seamless and ubiquitous interconnection of diverse environments and smart infrastructures. In such hybrid, complex and convoluted systems, agents acting on our behalf make the best decisions, for example finding the best deal on a travel package or making us aware of opportunities and available choices in a plethora of areas, such as "discovering" that there is no milk left in the fridge and ordering it, suggesting changing to a more ecologically friendly energy supplier, managing financial systems or materials and supplies in storage, or finding the best partners to help a company get that highly demanded product on the market sooner than the competitor can manage.

Designed following the natural laws of evolution, which merge self-organization and natural selection, these socially embedded information infrastructures can adapt to fulfil various needs as their environment demands, enabling the sharing of information, services and applications among suppliers, employees, partners and customers via:

- deployment of automated, intelligent software services (e.g., internet-enabled negotiations, financial transactions, advertising and bidding, order placement and delivery, automatic order tracking and reporting, etc.);

- complex interactions between such services (e.g., compliance policies, argumentation and persuasion via complex conversation protocols, bargaining, etc.);

- dynamic discovery and composition of services to create new compound value-added services (e.g., dynamic virtual clustering of synergetic partnerships of collaborative organizations aiming to achieve a common goal, finding and accessing an unknown service that is available on the Web, matching of different templates from different sources to design a product optimally, etc.).

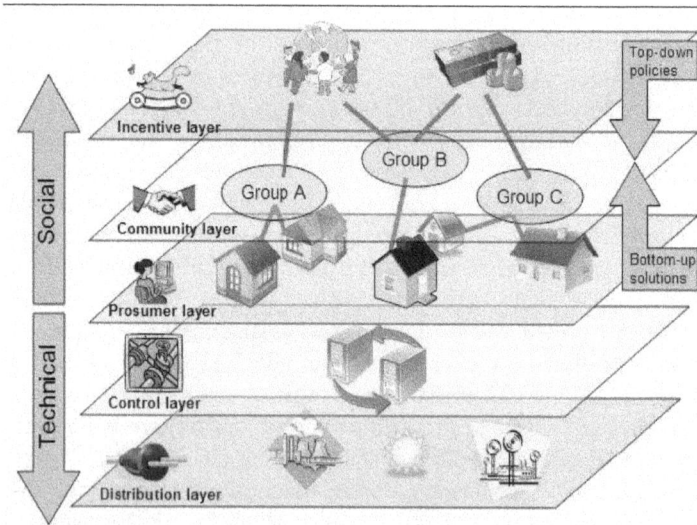

Figure 1. The EnergyWeb, integrating the cyber-physical and social dimensions.

Such digital ecosystems are envisioned also as an enabling technology for the "backbone" and "central nervous system" of the future smart infrastructures (e.g., Smart Grids, Smart Transportation Systems and Vehicles, Smart Buildings, Smart City) that will craft organic sustainable regions and enable the shift to a green and environmentally sustainable economy in low-carbon cities of the 21st century. In particular the Smart Grid is expected to have robustness, adaptability, self-healing and self-protective capabilities to support highly dynamical networks of power producers and consumers ("prosumers") through advanced ICT infrastructures, incorporating into various interdependent critical infrastructures the benefits of distributed computing and communications to deliver real-time information and enable the near-instantaneous balance of supply and demand. Along these lines the EnergyWeb concept (see Figure 1) is envisioned as a multi-layered large-scale socio-technical system in which industries, cities, communities or individuals will become part of a global socio-ICT "ecology," in which they can negotiate the energy they produce and consume. They will obtain direct financial benefits while promoting at the same time the growth of renewable energy sources. Since energy consumption by users and energy production by renewable energy sources are by nature unpredictable, utilization of the energy produced can be optimized by applying the idea of self-organization at the control level, influenced by the social network resulting from real-time involvement of prosumer communities in the operation of the grid.

Living digital ecosystems also offer a unique opportunity to address the crucial challenges related to the global nature of the Internet, which exacerbate the already critical global security risks facing our world. Besides the potential conflicts stemming from global warming, the possibility of "cyber war" adds an additional risk, threatening to destabilize interdependent critical infrastructures from finances and logistics to electrical and transportation grids on a large scale. The weaving of ICT into every critical infrastructure at a global scale across regions, countries and continents tremendously exacerbates the risks threatening business continuity and public safety, making risk and uncertainty even more integral parts of our lives. At the same time this becomes an opportunity to inject resilience and agility into critical infrastructures by designing the future networks with embedded autonomic self-* properties. Our "design for resilience" paradigm is using the ICT backbone endowed with self-healing, anticipatory and recovery capabilities; it presents a unique perspective on the implementation of global security strategies via a holistic approach which makes the security and trust of the ICT-infused interdependent critical infrastructures an integral dimension of public security.

4. What do you think are the most important open research questions about living technology, and how you think they should be pursued?

In what we refer to as "digital ecology" theory and practice our research aims to understand and advance the interweaving of humans and ICT to create a world with social, physical, and cyber dimensions, enabling a kind of social network in which humans are not just "consumers" of data and computing applications. Actors in the social network operating within the new digital ecosystem are much more: They are producers, "players," and "inputs" whose interactions use the "invisible hand" of the market to steer complex, interdependent global-scale systems linking hybrid sectors of the economy and society. We are looking for the principles of management and engineering of these emerging complex large-scale systems that will infuse them with the ability to discover a variety of potential solutions in their repertoire, when confronted with a problem-rich environment.

Based on concepts having to do with distributed networked multi-agent systems with a variety of purpose-built capacities (e.g., sensing, reporting, acting, and collaborating), our quest belongs to the novel area of "emergent engineering," which aims at solutions that can be selected through an evolutionary adaptation process to produce progressively better (and continuously improving) solutions. Our long-term goal is to rephrase the classical concepts of engineering design and systems control respectively in terms of the evolvability and emergence found in natural systems, to propose a breakthrough approach to the architecture and control of future digital ecologies. The uniqueness of our approach stems from our regard for self-organization as a paradigm for designing, controlling, and understanding complex distributed systems, fundamentally challenging the traditional engineering school of thought in its core principles.

We believe that it is now essential to develop a "living technologies toolbox" of methods and techniques that address the architecture of digital ecologies and the control of smart infrastructures for a sustainable world spanning production, agriculture, defence, finance and the economy as a whole. As an example, for the deployment of the Smart Cities of the 21st century we can see immediate applications within three interwoven areas:

a. Green Cloud Computing concerns management systems for the dynamic and distributed allocation of computing load to cloud computing resources powered by carbon-free sources, e.g., hydrooper, wind farms, solar farms, and tidal power. Our objective is to address challenges that relate to scalability, management, service models, quality of service, ease of use, scope of applicability and relation to pervasive ICT infrastructure, business models, and power and carbon emissions.

b. Smart Power Grid concerns the development of a management system that integrates sensor and communication technologies with the power grid to produce a system that encompasses generation, transmission, and distribution as well as end users and appliances. By leveraging real-time information, the management system will provide higher efficiency, reliability, and flexibility. The central role of micro grids and plug-in hybrid electric vehicles, as well as smart meters, applications in the home, and incentives to influence consumption behaviour will be considered.

c. Intelligent Transportation Systems concerns the development of a large-scale, distributed mobile networking and cloud-computing architecture to network and harness the power of millions of cars and public transportation vehicles. We seek novel approaches to designing digital ecosystems for networked vehicle infrastructure that can manage the flow of people and goods in an integrated private and public transportation network, to produce dramatic improvements in performance and reductions in environmental impact. There is a very high need for large-scale management systems that monitor and control the flow of vehicles for efficient movement of people and goods in an integrated road and public transit system. This integrated system will include dynamic pricing mechanisms and networked vehicles, as well as the interactions that will result between smart grids and the intelligent transportation systems mediated by the digital ecology.

5. What do you consider to be the most interesting and important human or societal implications of research and development in living technology?

Taken together, interconnected grids of communication, electricity, and transport amount to what we call "organic infrastructures" whose integrated and reliable operation will undergird development of this century's energy-efficient and sustainable cities, hosting the institutions and technologies of transformed low-carbon economies. Living technologies will enable the deployment of such organic infrastructures for the transition to a postmanufacturing, innovation- and knowledgebased green economy and society. We envision that such constructive organic organizations designed as "living systems" merging power grids, cloud computing, smart buildings and transportation networks will advance the state of the art in the rigorous design of resilient and robust management systems for the control of resource usage in massive-scale complex systems, targeting a "clean" and "green" world. We envision that living technologies will be at the forefront in the deployment of digital ecosystems that promote energy efficiency and reduce carbon and greenhouse gas emissions in homes, offices, factories, cities, and entire urban regions; and in extremely large-scale distributed data centres that will support the next generation of cloud computing.

Above all else, however, the enormous potential stems from the lessons that we can transfer from life's successful principles to revolutionize our governance structures to reset the current dynamics of our world from its perilous (market-and-conflict-oriented) trajectory, onto a prosperous one that is sustainable and focused on human needs. Such a shift in governance is required for fuelling the generation and adoptio of innovation, in all sectors and at all levels of social, and institutional and organizational structures. There is an acute need to (re-)define new indicators of wealth and social well-being that will enable this critical paradigm shift from risk governance to resilience governance. Due to the lack of adequate policy frameworks, the obstacles in implementing innovative solutions at all levels are the limited capacity of social processes to manage rapid change in institutional design, planning and public services, rather than technological innovation. Lessons from life's processes are priceless in restructuring the organization of our social processes into more fluid and organic structures that enable the manifestation of creativity through social innovation generation. Among the most critical issues that must be addressed without delay are:

- how to facilitate the *transition* from the currently disabling rigid governance structures into the necessary enabling agile policy frameworks that will transform our coercive institutional frameworks into agile, responsive and fluid ones, capable of fostering creativity and supporting innovation;

- how to design *engaging control mechanisms* that stimulate rather than oblige, transitioning the current work organization processes from contract to commitment by fuelling performance through visceral engaging architectures of participation, which, in an online gaming-like manner, infuse blissful productivity into work activities, giving an epic-like meaning to the purpose of work;

- how to design *validation frameworks* that reveal the impact of policies on the work ethics, culture and productivity in our organizations;

- how to redefine *indicators* that expose the impact of the convoluted effects of interdependent socio-political-economic factors on the current global dynamics, negatively affecting the overall wellbeing and sustainability of life on Earth;

- how to anticipate the evolution of society and the course of life under the influence of the transformative forces that change us as individuals – who in turn change our environment, which changes us – on the ever-mysterious trajectory of mankind's destiny as part of our "living," self-organizing universe.

Abouth the Author: Professor Mihaela Ulieru has held the Canada Research Chair in Adaptive Information Infrastructures for the e-Society since 2005. She also established (with the Canada Foundation for Innovation Funding) and leads the Adaptive Risk Management Laboratory (ARM Lab), researching complex networks as control paradigms for complex systems to develop evolvable architectures for resilient e-networked applications and holistic security ecosystems. She was recently appointed to Canada's Science, Technology and Innovation Council by the Minister of Industry, to advise the government and provide foresight on innovation issues related to ICT impact on Canada's economic development and social well-being against international standards of excellence. Professor Ulieru has a PhD in Diagnostics and Controls of Dynamical Systems from Darmstadt University of Technology in Germany and was on the Faculty at Brunel University in London, UK and at the University of Calgary in Canada where she held the Junior Nortel Chair in Intelligent Manufacturing and founded the Emergent Information Systems Lab. She is a highly respected expert in distributed intelligent systems, a topic on which she is a frequent keynote and tutorial speaker as well as distinguished visiting professor internationally. She has held and holds appointments on several international S&T advisory boards and review panels. To capitalize on her achievements and expertise and to make information technologies an integrated component of policymaking targeting a safe, sustainable and innovation-driven world, she recently founded the IMPACT (Innovation Management and Policy Accelerated by Communication Technologies) Institute for the Digital Economy, for which she currently acts as President.

19

Rinie van Est

Research Coordinator and Trendcatcher

Rathenau Institute's Technology Assessment division

1. In what sense do you find it meaningful to talk about "living technology?"

Up to now engineering was about building artefacts from nonliving material or influencing living organisms, for example through breeding. Today's technosciences are loaded with a higher ambition: to design and build artefacts and systems with lifelike features, like the powers to self-repair, grow and reproduce, adapt and evolve, and show emotion or make complex decisions. Long-term ambitions range from building living cells from scratch to designing autonomous robots. Short term ambitions include creating self-healing materials or developing cameras that can detect aggressive behaviour. The term 'living technology' captures this broad engineering trend in a clear and vivid way. Hopefully the term 'living technology' can help in making visible to a larger audience this engineering megatrend and thereby exposing to public scrutiny current developments in science and technology, and the visions driving them.

NBIC convergence
The wish to engineer lifelike artefacts has a long tradition, but got new meaning and impetus around the start of the 21st century, when the concept of NBIC convergence was launched in 2001 during the NSF workshop "Converging Technologies and Improving Human Performance." NBIC refers to four key technologies: nanotechnology, biology, information technology and cognitive sciences. Convergence refers to the belief that scientific and technological progress depends increasingly on the mutual interplay between those four key technologies. NBIC convergence was thought

to be crucial for the successful development of new areas like molecular medicine, service robotics, ambient intelligence, personal genomics and synthetic biology. Accordingly, NBIC convergence created a new set of engineering ambitions with regards to biological and cognitive processes. If these ambitions would only partly come true, a new technology wave was to be expected; even a new phase within the ongoing information revolution. To position "living technology" within this new technology wave, it is instructive to look at NBIC convergence as consisting of two complementary megatrends. In this respect, W. Brian Arthur says that "biology is becoming technology" and "technology is becoming biology." To me "living technology" refers to the latter, very relevant but rather hidden megatrend in science and engineering.

Biology is becoming technology
Society is quite familiar with the first megatrend, "biology is becoming technology," which refers to the steady growth in the set of engineering tools for studying, modifying and copying (parts of) living organisms. In particular, since the 1990s, society has become acquainted with this trend through a wide range of heated debates around GM food, cloning, xeno-transplantation, embryonic stem cells, embryo selection, etc. Themes like "messing with nature," "playing God" and the "instrumentalisation of life" play a central role within these debates. Within the frame of NBIC convergence, nanotechnology and information technology are further enabling technological progress in the life sciences, including the cognitive sciences. This broadens the bio-political debate in various ways. Besides gene technology, information technology provides means to intervene in living organisms. Think about smart electronic pills to monitor disease or deliver drugs in the human body, brain pacemakers and smart prosthetics. The debate also broadens out from micro-organisms, plants and animals towards humans. NBIC convergence has already led to a growing international debate on human enhancement, i.e., the promises and perils of engineering the human body and mind.

Technology is becoming biology
As I said, I take the words 'living technology' to refer directly to the second complementary trend: "technology is becoming biology." This trend is as relevant from a social point of view as the former trend, but is much less reflected upon. First of all, the concept "living technology" has relevance for understanding the stage science and technology are currently in and uncovering

important visions that drive engineers. For example, the ambition to build artificial cells from scratch, so-called protocells, a research goal that originates from the small artificial life (ALife) community, has evolved into one of the central challenges for biochemists in the 21st century. They are dreaming about controlling the most complex interaction between molecules: that of spontaneous reproduction. (In fact, the term 'living technology' stems from the ALife community, and it is no coincidence that protocells are often used as a prime example of living technology.) The goal of designing autonomous robots also has become part of today's mainstream science. Robot experts around the globe have set up RoboCup – the annual world soccer championships for robots – aiming to build a team of cooperating humanoid robots that is able to beat the human world champion in 2050. The American army is even building a robot soldier that is assumed to make more "human" decisions about life and death in the heat of the battle. Even the dream of nanobots swimming around in our bodies and fighting diseases like cancer has entered the realm of science. Although these bold targets may never be achieved, they get a lot of media attention, because they trigger our (moral) imagination. These high-profile projects and visions, however, present only the proverbial "tip of the iceberg." It would be a real pity if we were to dismiss the part of the iceberg that is under the surface of the water, because it presents a new way of doing science and technology, which will have a major impact on society and the way we interact with nature.

Biomimetics

While nanotechnology and information technology enable "biology becoming technology," "technology becoming biology" is driven by convergence in the opposite direction. Here the life sciences – insights in biological and cognitive processes – inspire and enable progress within the material sciences and information technology. This development relies heavily on so-called biomimicry or biomimetics. The basic idea behind biomimetics is that engineers can learn a lot from nature. Namely, from an engineering perspective the evolution of life might be seen as a gigantic billion-years "R&D experiment," which through natural selection has led to elegant and highly efficient biological, neurological and social mechanisms and "machines" that may inform innovative solutions to many current environmental, health and social problems. Engineers want to emulate nature to enhance their engineering capabilities. In this line of thinking, algae may provide a bio-solar system that is more efficient than the silicon-based solar cells our engineers have created. But although nature's achievement is impressive, engineers think that there is still plenty of room for improving the engineering skills of nature. For example, algae absorb blue and red light, but not a lot of green light. Engineers would like to design bio-solar systems that can do it all.

On the molecular level, nanoscientists and biologists alike have started to view the living cell as a factory crowded with numerous nanomachines that engineers should be able to mimic. Scientists all over the world are drawing inspiration from biological mechanisms to make complex molecular systems, for example, molecules that can convert light into rotary movement. Engineers also take inspiration from the brain; for example, for developing new techniques for analysing medical images that may help or even one day outperform doctors. The Swiss Blue Brain project presents another example of brain-mimicry. The engineers involved are building a computer model of the brain of a hamster, based on physiological measurements. The aim is to get a better understanding of the brain and brain disorders. But this way of re-engineering the brain down to the molecular level may also contribute to the field of artificial intelligence (AI). Engineers are also working on technology that can detect and mimic social behaviour. For example, software has been developed to detect the reaction in people's faces when trying on new cloths and give feedback accordingly. All these examples show that our technological capability and level of understanding enable engineers to go beyond the "simple" mimicking of dead nature, and make a bold step in the direction of biological-, neurological-, social- and emotional-inspired approaches towards science and engineering.

"Living technology" as a sensitizing concept
"Living technology" accurately catches this engineering megatrend. I hope the words 'living technology' may serve as a sensitizing concept, a conceptual lens, which allows people to get a better and more comprehensive view of these developments in science and technology. Only by putting this development on the radar, will it be possible to start reflecting upon the social meaning of this development and discussing that in public.

2. How does your research relate to living technology, and why were you initially drawn to this work?

My basic interest lies in the politics of innovation. My PhD thesis *Winds of Change* (1999) compares the politics of wind energy

innovation in Denmark and California. One finding was that political traditions and ways of innovating are closely related. The
arrival of democracy and political ideologies, like liberalism and
socialism, for example, coincided with and shaped the industrial
revolution. The stimulation of wind energy was a particular political response to another huge societal problem associated with
industrialization, namely the environmental crisis.

Since 1997, I've worked at the Rathenau Institute, the Dutch
national technology assessment organisation. Its task is to stimulate public debate on the social and ethical aspects of science and
technology, and inform the Dutch parliament about that. During
those thirteen years, I have had the privilege and opportunity to
look at a very broad spectrum of new emerging technologies and
debates, ranging from cloning, genomics, nanotechnology, brain
sciences, ambient intelligence, converging technologies, human enhancement, robotics and synthetic biology. I have always looked
for connections between these developments, and discovered common ground. One of the connections is that all these fields are
typical for our current information-based society. While industrial
technologies are aimed at controlling the "nature" surrounding us,
the above technologies focus on our memory and personality, human reproduction, physical achievements and social relationships.
In other words, our modern information technologies aim at controlling the fundamentals of life and human consciousness. I am
deeply interested in what way the information revolution is creating a new politics and technological culture.

My current function at the Rathenau Institute is research coordinator and trendcatcher. At our institute, I have often played a
role in signalling new developments, putting them on our agenda
and setting up new related projects. The title "trendcatcher"
refers to that agenda-setting role and formalizes it. It also acknowl-
edges that technology assessment can be more than assessing the
impacts of one specific technology, like radio frequency identifica-
tion tags or gene doping. It is also important for technology assessment to signal broader trends and raising awareness about the
social and cultural meaning of such socio-technical trends. In this
respect, nanotechnology and NBIC convergence have intrigued me
from the very start, because they bring to the eye a whole range
of new developments within science and technology, like molecular
medicine and personal genomics. NBIC convergence presents an
ideal frame for catching trends in the modern technosciences.

The way in which the above mentioned NSF workshop framed

NBIC convergence also generated a personal drive. Linking NBIC convergence with the goal of enhancing humans in order to give individuals greater opportunities to achieve personal goals revealed a clear political bias towards libertarianism. Driven by democratic ideals, I wanted to contribute to raising political and public awareness about this new technological wave. My aim was to enable a politically more pluralistic debate on the social meaning of NBIC convergence. One line of action was to signal the technological utopianism behind NBIC convergence. Traditionally technology is about building nonliving artefacts, like cars and bridges. Still, we are accustomed to re-making the living world with the aid of technology. For example, we use biotechnology to genetically modify E. coli bacteria in order to produce insulin to treat diabetics. This involves technical ability to make things, but it is a different type of "making" from when we build a drilling machine or power grid. Up to now, we *build* nonliving matter; we *influence* organic life: micro-organisms, plants, animals or humans. NBIC convergence epitomized a radical expansion of the building logic of nonliving nature in the direction of living nature (Swierstra et al., 2009).

This basic insight made me ponder the way engineering in the information society would deal with and change (human) nature. To come to terms with this issue, I started to write an essay with the working title "Our second nature." The essay was meant to conceptualize the content of a technology festival, which the Rathenau Institute used to organize every two years. This title refers to the fact that using technology is our natural inclination and that this habit has enormous effects on our environment. The agricultural and industrial revolutions have created manufactured landscapes; even our climate has become man-made. At the time, developments like synthetic biology, Second Life, social robotics, and the debate around human enhancement all pointed at ways in which the current information revolution is creating new types of synthetic worlds.

Unfortunately, the organisation of the technology festival has been cancelled, and as a result the essay has never been finished. Nevertheless, this unfinished work formed the starting point for setting up the project "Making perfect life: Bio-engineering (in) the 21st century" (van Est et al., 2010), which is commissioned by the STOA-panel of the European Parliament (STOA stands for Scientific Technology Options Assessment). Moreover, it made me accept the invitation to participate in the "Living Technology Project," which was set up by Odense University as part of the Initiative for Science, Society and Policy (ISSP). I hoped at the time that getting involved in the debate on living technology would help me elaborate my thoughts about the social and cultural meaning of NBIC convergence and the politics of the information revolution.

3a. How is living technology related to overlapping or nearby areas, such as nanotechnology, molecular biology, cloning and stem cell research, genetic engineering and synthetic biology?

I believe that the representation of life as a digital control system, which was formulated at the beginning of the information revolution, is the thread that connects living technology with all the other areas.

Digitisation of life
Life and technology refer to totally different worlds. Maybe life is the part of nature most distinct from technology. "Living technology" implies an attempt to cross this fundamental and emotional dividing line between life and technology. It aims to bring life into the domain of technology and engineering. Scientists and engineers have a trick to do that: defining life in technological terms. Science has a long tradition of doing that with great consequences for man's relationship with nature. Descartes looked at living organisms in mechanistic and static terms. Mechanisation of life has (had) major consequences for the way human beings justify their control over nature. The engineers of our information age are guided by another concept of life: *life as an information processing system.*

This new scientific understanding of and outlook on life was pioneered within military research during the Second World War. Two central pioneers who pondered the similarities between life and machines were John von Neumann and Norbert Wiener. During the war von Neumann had contributed to designing computer hardware and software for the Manhattan Project – building the atomic bomb at Los Alamos. In 1948 he posed the question of complexity and bottom-up self-organization. His investigation of self-reproducing machines inspired scientists who were dealing with genetics at the time. Based on his work in the field of automatic gunfire control, Norbert Wiener conceived of cybernetics, the use of information theory to design self-correcting machinery.

In 1948, he published *Cybernetics: Or Control and Communication in the Animal and Machine.* Wiener described animals and machines alike as information processing systems that constantly move from action based on a certain goal, to sensing the environment, to comparing those results with the desired goal, and then acting on that feedback accordingly. As a consequence Wiener held that like animals, machines could be built to behave purposefully. Science had managed to redefine life – biological, cognitive, and social processes alike – in terms of automated digital control, in a language that belonged to the realm of engineering. This cybernetic view of life presented a new scientific way of looking at the world and shaping it. The digital control paradigm laid the symbolic basis for engineering the information revolution.

Molecular biology and genetic engineering
During the 1950s, many disciplines within the social and life sciences became inspired by cybernetics, systems theory, information theory and computers. Also molecular biology – which was based on the belief that it should be possible to build a new biology on the basis of the physical sciences – began to represent itself more and more as an information science. This discursive shift got impetus when in 1953, Francis Crick and James Watson unravelled the double-helix structure of DNA, proclaiming to have found "The secret of life." This date is often seen as the birth of molecular biology. But because the genetic code became signified as a (programmable) information system, it also meant the breakthrough of the computational view on life.

The development of recombinant-DNA (rDNA) technology really opened up the possibility of reprogramming living organisms. In 1973 genetic engineering was accomplished in micro-organisms. rDNA technology enabled scientists to cut specific parts of the DNA, and paste them into other cells. Scientists could suddenly combine the genes of two separate species of living beings, and thus consciously and purposefully create new forms of life. One of the big achievements of this first generation gene technology was the genetic modification of an E. coli bacterium in order to produce human insulin for diabetics. The rDNA technology also led to discussions about biosafety. For example, in 1975 the famous Asilomar Conference on rDNA was held to draw up voluntary guidelines to ensure the safety of this technology. At the time little attention was given to ethical issues.

Systems biology and artificial life

The potential that lies in the merger of the digital and biological worlds was fully captured in the 1980s through two visions: systems biology and artificial life. Both strands enabled and constituted the arrival of the field of synthetic biology at the beginning of this century. The systems biology vision was pioneered by Leroy Hood at the CalTech biology department. He imagined machines that could sequence and synthesize DNA molecules and proteins, and believed that based on a complete "parts list" a computer model of a living system could be built. Hood's ideas about merging the digital and biological worlds have been very influential. The 1990s saw the Human Genome Project, which depended heavily on bio-informatics. Hood's vision of digitizing biological processes in the cell has nowadays become the central tenet of molecular medicine, which aims at studying and treating diseases at the cellular and molecular levels.

The more marginal artificial life vision developed within the computer sciences. The ALife community aimed to create artificial systems that exhibit behaviour characteristic of natural living systems. ALife was directly based on von Neumann's theory of self-reproducing cellular automata. The 1970s had already shown some forms of artificial (digital) life, namely the Creeper computer virus, a self-replicating program on ARPANET, which distributed the message "I'm the creeper, catch me if you can!" By the 1980s, AI scientists found they could simulate life and evolution as well as intelligence and experience. This new research branch was called Artificial Life, and was centred in Los Alamos. It was expected that IT and genetic engineering would soon enable the creation of new life forms *in silico* as well as *in vitro*. For ALife life is not bounded to organic, carbon-based and DNA-based life forms, nor to any type of substance. Instead, life is interpreted as "a model of the interconnectedness, emergence and behaviour of the constituent components of a(ny) living system" (Parikka, 2005).

Synthetic biology
Synthetic biology is a typical recent form of digital biology, which includes the ambition to build and program from scratch living artefacts. The ambition to create synthetic life is not new. Already at the beginning of the 20th century, the German-American physiologist Jacques Loeb (1859-1924) dreamed of "a technology of the living substance." Technological progress has revived this ambition, big time. Synthetic biology builds on the idea of convergence among biotechnology, information technology and nanotechnology. It presents a new type of bio-futurism fuelled by the

success of the Human Genome Project. While genomics was about mapping and understanding biological systems, synthetic biology aims at designing and constructing new, complex artificial forms of life.

We may distinguish between a "top-down" and a "bottom-up" approach to synthetic biology. The "top-down" approach starts with a pre-existing living system and then re-engineers it for some purpose. Various strands of research can be distinguished. The well-known research area of metabolic engineering has reached a new level of complexity, being able to design genetic circuits of about 10 to 15 genes. Critics talk about extreme genetic engineering. The great challenge, however, is to standardize bioengineering. The idea is to create standard biological parts (so-called BioBricks) to enable bioengineers to create genetic circuits with the same ease with which electronic engineers build computers systems. The "bottom-up" approach attempts to make simple kinds of chemical cellular life, using nonliving material; so-called protocells. This latter approach has its roots in ALife.

I do not believe that counting a technology as a living technology is related to the engineering approach taken to construct it. Instead, the trends "biology becoming technology" and "technology becoming biology" can close in on each other and start to intermingle. Take for example Craig Venter's (top-down) effort to create a cell with a minimal genome, which is meant to be used as a cellular platform for engineered biological platforms. The Craig Venter Institute has already successfully engineered the first bacterium with an artificially synthesized genome, and thus falls into this category. Also, Venter talks about this research in informational terms, like "booting up the synthesized DNA into a living cell," and "synthetic DNA as software that builds its own hardware." On the YouTube film "Creating artificial life"[1] Venter argues: "We don't have to design life from scratch, we just have to design the software appropriately." If an engineer is able to control a living organism – whether built from scratch or fleeced – to a certain high level, such biology has become a sort of technology and thus we may speak of living technology.

Cloning and stem cell research
For cloning and the use of stem cells a similar argument can be made. Cloning may be used for reproductive or therapeutic pur-

---

[1] http://www.youtube.com/watch?v=iQ1VNEgcWE8

poses. The announcement of the birth of Dolly in *Nature* in 1997 led to a world-wide debate in particular about the question of whether reproductive cloning of humans would become technically possible in the future and whether this would be morally acceptable. In therapeutic cloning, stem cells are created to regenerate tissue and organs. It is guided by the hope to circumvent the problem of immune rejection. From an engineering point of view and in theory, cloning is the ultimate reproduction technology, because it creates a genetic copy of (a cell of) an existing living organism, whether natural or genetically engineered. Human cloning and therapeutic cloning are both experimental technologies. In the hypothetical case that engineers could totally master, for example, stem cell therapy, I would designate stem cells as a living technology. In that situation "biology would have become technology" and vice versa.

3b. How is living technology related to social and technological systems such as social networks or information networks, such as the world wide web, cell phone networks and electronic banking networks?

Entangled social and technological systems
In particular among engineers and natural scientists the idea that technology development follows its own dynamics, and is *the* dominant driver behind social and cultural change is still very much alive. Within this technological determinist view, posing the question of the purpose of the World Wide Web and cell phone networks would be senseless. The most basic insight from science and technology studies (STS), however, is that technology and society co-construct each other. Therefore the development of a technological system is inherently a social process, guided by the visions, goals, actions and interactions between many different social groups: from engineers and consumers to policy makers. Large technological systems "work" or "don't work" not only because of the interplay between the various technological components, but also because of the efforts of people who are guided by ethical norms, laws, incentive structures, negotiated technical standards, etc. For example, it was the users who unexpectedly embraced SMS (short message service) technology and changed the Internet from an information to a social web. In turn, social media have an enormous influence on how people interact and organize their lives.

Social choice, politics and technology are thus entangled. This implies that the engineering megatrend "living technology" is a man-made socio-technological trend. From a social constructivist point of view each technological system can be regarded as "living," in the sense that each socio-technical system is a hybrid between technological and social (living) systems. However, as I clarified under question 3a, the "living" in "living technology" does not relate to a social constructivist view on technology, but to an engineering perspective on life, in which life is defined in technological terms. To avoid serious conceptual confusion, I will stick to the engineering interpretation of "living technology." This implies that I will not further discuss the relation between living technology and social networks and systems. My position on that issue has been explained above; they form a so-called seamless web. But I will come back to the utterly relevant question of how the (living) technological, social and natural worlds interact with each other at the end of this interview. In the remainder of this section, I will focus on the extent to which "living technologies" have become part of the technical engineering of socio-technical systems, like the World Wide Web, cell phone networks, and electronic banking networks.

Early visions of the information society
At the start of the information age, the ambitions and visions connected to the arrival of the digital control paradigm were very high, sometimes even divine. The computer itself was meant as a machine to imitate and in the end transcend the human mind. From the start the computer was meant to be a sort of "living technology" in the sense that it could show "intelligent" behaviour, which at the time was equated with formal logic. Some foresaw the physical integration of man and machine, the cyborg, a fusion of living organisms and cybernetic systems. Some saw the universe itself as a gigantic computer, as God's simulation. The NBIC convergence enterprise – basically modern science – is still guided by the same digital control model of life and machines. To illustrate this, participants of the already mentioned NSF workshop "Converging Technologies for Improving Human Performance" expected around 2020: fast broadband interfaces between the human brain and machines, the ability to control human genetics, and networks of smart wearable sensors that constantly monitor the health and environmental conditions of plants, animals and human beings. It was even prophesized that at the end of the 21st century humanity would have evolved into "a single, distributed and interconnected "brain;" one huge, integrated cybernetic "liv-

ing" system.

Half a century into the information revolution, the computational paradigm has penetrated and radically transformed many industrial sectors and social domains. Driven by the long-term exponential development of computer technology engineers have put a digital layer over production, transportation, communication, the globe, public space and life itself. Local and global networks of computers have become ubiquitous. Numerous remarkable engineering successes can be named, like GPS, the automatic pilot to navigate airplanes, missile defence systems, and putting a man on the moon. Information technology has been responsible for a wave of breakthroughs in many areas, such as novel materials, energy and transport systems, medical applications, and robotics. Computer technology has also revolutionized our way of interacting and communicating. Just think about the arrival of the World Wide Web during the 1990s, and the current rapid integration between Internet and mobile telephony. But to what extent do these information networks relate to the concept of "living technology?" I believe this concept tells us something about where these information technologies are going. The following examples show the contours of this trend.

Delegating more complex tasks to computers
NASA's website presents its spacecraft Deep Space as "one small step in the history of space flight, but one giant leap for computerkind." What makes Deep Space so special that NASA paraphrases Neil Armstrong's famous words when he made his first step on the moon? Former space flights had preprogrammed flight schedules and were ground controlled. When at the end of 1990s Deep Space was some 60 million miles away from Earth, the primary control of the spacecraft was handed over completely to an artificial intelligence system. Also on Earth steps towards delegating control towards intelligent machines can be witnessed in various fields. Unmanned airplanes, like the *Predator*, can remain airborne for twenty-four hours, and are currently employed extensively in Afghanistan to trace Taliban insurgents. These drones can fire missiles and are flown by pilots – so-called cubicle warriors – who are located at a military base in the Nevada desert, thousands of miles away from the battlefield. Modern networked cars may include technological features, like intelligent breaking systems and automatic cruise control, which depending on the traffic and speed regulations makes the car slow down or speed up, or systems which warn drivers when their car leaves its lane, and take action in case

drivers do not react. There are also network navigation systems, which base their information on the mobile cell phone use of other drivers and inform those drivers about current traffic jams.

The virtual world presents similar examples of delegating (part of the) complex decision-making to computers. If we buy a book on eBay, computer programs provide immediate "personalized" advice about products we might also be interested in. These ads are based on the shopping behaviour of people who in the past have shown an interest in the same book. This type of advice – or psychological therapy and pedagogical support during your study – might be given by a computer-generated character, showing human interest and emotion. High-frequency trading even takes people out of the control loop. This type of robo-trading currently accounts for a huge share of the world's equity trading volume. Here autonomous computer algorithms make trading investment decisions, which involve huge sums of money, within less than a millisecond.

Digital control has entered a new phase
All these examples show that digital control has entered a new phase. This broad development is signalled by myriad different headings, like ambient intelligence, domotics, intelligent cars, service and social robotics, affective and persuasive computing, ubiquitous network society, from the Internet to robotics, etc. The common thread is a new engineering vision on intelligence. For a long time, the AI community has looked for ways to adequately model the world via symbols and logical inference. This approach had some successes – like the chess-playing computer Deep Blue who beat the world champion Kasparov in 1997 – but has failed in dealing with many other control issues. Over the last decades, this rationalistic, abstract "top-down" modelling approach has been complemented with a new "bottom-up" perspective on intelligence, which pays attention to the roles the body, emotion, and dynamic environments play in intelligent behaviour. The aim is to engineer goal-directed machines (virtual or real) that are able to explore and interact in a dynamic way with complex physical and social environments (virtual or real), and can recognize and seize opportunities. For example, the US army has developed an Energetically Autonomous Tactical Robot (EATR) that is powered by biomass. The robot can obtain its energy by foraging, or in engineering terms "by engaging in biologically-inspired, organism-like, energy-harvesting behavior." Inspired by the humanities, biology and cognitive sciences, engineers thus want to build technology

that shows problem solving on the spot and can cooperate with other machines to reach a certain goal. In this case, the 'living' in 'living technology' thus refers to a new engineering vision on (artificial) intelligence. Exponential decrease of the cost and size of computer power, combined with developments in sensor technology and programming (in particular agent-based modelling software) enables and legitimizes this new AI research path.

4. What do you think are the most important open research questions about living technology, and how do you think they should be pursued?

5. What do you consider to be the most interesting and important human or societal implications of research and development in living technology?

For a technology assessment practitioner, Question 5 points at the most relevant research questions. Accordingly, from my perspective Questions 4 and 5 are pretty similar, and I will treat them as one and the same question. I would like to rephrase Question 5 in a somewhat more open way though. Namely, "human and societal *implications* of R&D" hints to technological determinism. Taking into account the fact that society and technology co-evolve, I prefer to talk about and investigate the social and cultural *meaning* of living technology, i.e., the engineering megatrend of "technology becoming biology." Below I formulate some research themes that are worthy of further investigation, exploration and public and political debate.

Hybridization of the technological, social and natural world
Living technology provides us with a vast technological terrain that is in need of assessment from a social point of view. So where and how to start such an exercise? One of the central questions that needs to be addressed is how the "living" technological world will interact with the social and natural worlds. We have experience with the way "nonliving" technologies from the industrial age have interacted with the social and natural worlds. Let us look at two examples: plastics and cars. The automobile has given us freedom of mobility, its use has changed our outlook on our cities, roads can connect far-away families, but also split up local communities. Our love of cars has made us dependent on oil, and the way we use cars kills thousands of people and animals each year, through traffic accidents, but also through pollution. (Semi-)synthetic plastics have come to imitate, improve on and replace

traditional, more "natural" materials, like leather, wool, wood, marble, paper, glass, metal, etc. Due to their ease of manufacture, low cost, and imperviousness to water, plastics are used in an enormous range of products, like toys, bottles, jewelry, and computers. As a consequence of the way we use plastics, tiny plastic particles can be found in every location and in each living creature on our planet. Only recently it has been discovered that huge parts of our oceans are polluted with tiny plastic particles, so-called plastic soup.

Both examples show the ever-evolving complex relationship between the technical, social and natural worlds. The two stories remind us – in the words of Edward Teller – that "things b*ite back,*" *and sometimes brutally. Introducing technology in society leads to a lot of unintended effects, positive and negative. Properly dealing with these social consequences of technology* requires a lot of human vigilance and effort. It is important to try to anticipate possible drawbacks of the logic implicated in technology in an early stage of development. The logic of "living technology" introduces an extra complicating factor. While engineers traditionally design technology to perform a certain specific function, "living technology" is engineered to have, at least partly, a "life" of its own. The engineers of "living technology" are thus purposefully building in unforeseen effects and unpredictable behaviour. This unfamiliar conception of engineering raises special regulatory, but also ethical and cultural concerns. Important research themes thus are: How may we understand the engineering logic implicated in living technology? How will the megatrend of "technology becoming biology" impact the hybridization between the technological, social and natural world? How can society anticipate these sociotechnical changes?

Six types of living technologies
Cars and plastics illustrated respectively the mechanical machines and synthetic materials of the industrial revolution. The digital control paradigm of the information revolution guides the development of "living technology." The examples of living technologies given above hint at the existence of several types of living technologies. At least the following six types of living technologies are currently under construction: molecular machines, artificial cells, synthetic organisms, virtual worlds, smart environments and robotic machines. An important research question is how will these various (semi-)synthetic living environments interact with the social and natural world? I will explore the significance and scope of

this question by first looking at molecular machines and the way this type of living technology may interact with the natural world. Then I will take a glance at robotic machines and virtual worlds, and focus on how these technologies with lifelike characteristics interact with the social world. A central issue is to what extent these (semi-) synthetic worlds will conform to the standards set by the social and natural world, or whether they go beyond that.

Domesticating molecules
One response to the environmental problems plastics are causing is developing biodegradable plastics. Engineers are thus challenged to build in a characteristic in synthetic materials which all natural organic materials share. Organic materials show an important characteristic of life, namely, decay and mortality. Engineers often focus on the active features of life, like adaption and reproduction, but seem to have a blind spot for life's transient features, like sleep, living with the seasons, forgetting and death. With respect to molecular nanomachines a similar design question arrives. Should engineers try to mimic and learn from existing natural nanomachines, or should they develop new types of nanosystems? Mimicking existing biotic nature would be a sophisticated manner to connect technology and nature. In that case, nature presents the golden standard. The second approach would create a synthetic world on the molecular level separate from nature but of course interacting with it. Engineers seldom feel bound to what nature provides. A quote from the Dutch nanoscientist Ben Feringa illustrates this neatly:

When building molecules, we like to study the fantastic designs that nature has come up with. However, evolution has selected an extremely limited number of building blocks – primarily carbon and hydrogen – for all essential functions. Many substances are unusable, if not downright toxic, in living organisms. But that doesn't stop a chemist from using them. We're not bound to the same restrictions. We can drive our nanocar over a gold strip if that suits the particular properties of our molecular system. When it comes to molecular building bricks, we have a huge toy box and an unlimited playground. (quoted in van Santen et al., 2006, p. 214)

If we are going to domesticate the molecular world and build all kinds of artificial molecular machines we had better anticipate the "unforeseen" creation of "artificial molecular soups" popping up in our living cells in 2050. In other words, it is important to start reflecting on the environmental limits to the molecular engineering playground.

Dealing with robot agency

How will humans interact with the world of robotic machines we are building? Developments in the field of intelligent cars and military robots show that such interactions are multi-faced and hard to predict in advance. Above I sketched how the automobile is gradually changing from a (nonliving) machine into a machine stuffed with intelligent components; a robotic machine. One of the aims is to increase the safety of the car, but it is not yet clear to what extent this aim will be reached. Maybe the technological safety measures will even lead to more reckless driving styles. Besides hardware, these intelligent electronic systems are also dependent on software, thus introducing a new type of vulnerability. For example, Toyota recently recalled its Prius hybrid models because of problems with the software that controls the car's anti-lock braking system. Not only do things bite back, but humans may also undermine the effectiveness of technological systems. In a Dutch documentary film, a Taliban fighter explains that his group members always scatter around over a certain area, because they know that the drones were programmed to look for groups of Taliban insurgents. Moreover, *The Guardian* announced on December 17, 2009 that Iraqi insurgents had managed to hack the US drones, with cheap software available on the Internet, so they could view the potential targets of the drone.

(Semi-)autonomous robotic systems always raise the question of who is responsible if something goes wrong. This question will become more urgent if robots become more autonomous, but may still be relevant if semi-autonomous robot systems have a man-in-the-loop. For example, the drones that fly over Afghanistan and Iraq are operated by so-called cubicle warriors. In *Wired for War* a young pilot is quoted who operates drones over Afghanistan and Iraq, and describes how he experiences fighting from a cubicle: "It's like a video game. It can get a little bloodthirsty. But it's fucking cool" (Singer, 2009, pp. 308-309). For newly recruited soldiers, who have been playing videogames throughout their teenage years, there might not be a huge contrast between the experience of playing a video game and that of actually being a cubicle warrior. This socio-technical system seems to condition the cubicle warrior to dehumanize the enemy, which might morally disengage the cubicle warrior from his destructive and lethal actions (Royakkers and van Est, 2010). Consequently, it is doubtful whether the cubicle warrior can be held reasonably responsible for his actions. The question then becomes: Who is? This is highly relevant if one realizes that local authorities in Pakistan claim that near the Afghan border drone strikes on Al Qaeda and affiliate targets have killed more than six hundred civilians.

The modified reality of virtual worlds

The way people interact with each other and make decisions is increasingly mediated by virtual environments. This raises the question to what extent the virtual world simulates or manipulates the physical world, and thus to what extent it enables or distorts social human interaction and decision making? The above example of the drones illustrates how the virtual world impacts the interaction between soldiers and insurgents or civilians on the battle field. The decisions made by the cubicle warrior are strongly influenced by the fact that the robotic systems give the cubicle warrior a form of tele-presence on the battlefield, by the way a war scene is digitally represented on his computer screen, and the fact that he feels safe and in control so far away from the real killing fields. Digital worlds may lead to confusion about what is real and what is manipulated, because everything that has been digitally coded, can be copied and recoded quickly, cheaply and often unnoticed. The current political discussion in Israel about whether photoshopped pictures of models should be labelled is a good sign for things to come. Let me give one example of the future of interaction in the digital world. It is a well-known psychological finding that people judge people who resemble themselves to be more trustworthy. In the virtual world you are able to digitally mix features of another person's face into your own face. Through such digital modification of your face you might look more trustworthy to the person you are meeting. In this way, negotiations in a virtual world can be easily manipulated. Distinguishing between real and fake, between one-to-one modelling of the natural social world and digitally modified social worlds will be an important political, social and cultural topic for the future of digital control.

Concluding remark

Applying the computational paradigm of life on a massive scale to remake life and to build technology with lifelike characteristics asks for critical social and ethical reflection. As mentioned above, Descartes looked at life in mechanistic terms. His theory of animal-machines has legitimized the utilization of animals as tools by humans. Mechanisation of life had major consequences for the way human beings justified their control over nature. In

a similar vein, we need to reflect on how the digitisation of life –
and the digital machine metaphor of living organisms, including
humans – will lead to new ways of looking at life and controlling (human) nature. When "technology is becoming biology" the
distinction between life and technology, nature and culture, and
real and fake social interactions evaporates. Scientific reflection
and public and political discussion is, therefore, urgently needed
to start thinking about the social and cultural meaning of this
engineering megatrend.

About the Author: Rinie van Est is research coordinator and
trendcatcher with the Rathenau Institute's Technology Assessment division, which informs the Dutch parliament and stimulates
public debate on social and ethical issues related to science and
technology. Van Est studied applied physics at Eindhoven University of Technology and political science at the University of
Amsterdam. His PhD thesis "Winds of Change" (1999) examined
the interaction between politics, technology and economics in the
field of wind energy in California and Denmark. He joined the Rathenau Institute in 1997 and is primarily concerned with emerging
and converging technologies such as nanotechnology, cognitive sciences, persuasive technology, robotics, and synthetic biology. He
has many years of hands-on experience with designing and applying methods to involve experts, stakeholders and citizens in
debates on science and technology in society. Examples are the
Dutch debate on cloning and nanotechnology, and the European
public panel on brain sciences called Meeting of Minds. In addition
to his work for the Rathenau Institute, he lectures on Technology
Assessment and Foresight at the School of Innovation Sciences of
the Eindhoven University of Technology.

References

Arthur, W. B. (2009). *The nature of technology: What it is and
how it evolves.* London: Allen Lane.

Parikka, J. (2005). The universal viral machine: Bits, parasites
and the media ecology of network culture. Available online at:
http://www.ctheory.net/articles.aspx?id=500.

Royakkers, L., & van Est, R. (2010). The cubicle warrior: The
marionette of digitalized warfare. *Journal of Ethics and Information Technology.* DOI 10.1007/s10676-010-9240-8

Singer, P. W. (2009). *Wired for war: The robotics revolution and
conflict in the twenty-first century.* New York: The Penguin Press.

Swierstra, T., Boenink, M., Walhout, B., & Van Est, R. (Eds.) (2009). Special issue: Converging technologies, shifting boundaries. *Journal of Nanoethics*, 3(3), 213-216.

van Est, R., van Keulen, I., Geesink, I., & Schuijff, M. (2010). *Making perfect life: Bio-engineering (in) the 21st century. Interim study.* Brussels: European Parliament, STOA.

van Santen, R., Koe, D., & Vermeer, B. (2007). *The thinking pill: And other technology that will change your lives.* Amsterdam: Nieuw Amsterdam Uitgevers.

About the Editors

Mark A. Bedau (Ph.D. Philosophy, UC Berkeley, 1985) is Professor of Philosophy and Humanities at Reed College and a regular Visiting Professor at the European School of Molecular Medicine in Milan, Italy. He is an internationally recognized leader in the philosophical and scientific study of living systems and has published and lectured extensively on issues concerning emergence, evolution, life, mind, and the social and ethical implications of creating life from nonliving materials. He has published over 100 research papers and co-authored or co-edited 7 books, including *Emergence: Contemporary readings in philosophy and science* (MIT Press), *The prospect of protocells: Social and ethical implications of creating life from scratch* (MIT Press), and *The nature of life: Classical and contemporary perspectives from philosophy and science* (Cambridge University Press). He is Editor-in-Chief of the journal *Artificial Life* (published by MIT Press), co-founder of ProtoLife Inc., co-founder of the European Center for Living Technology (Venice, Italy), and co-founder and director of the Initiative for Science, Society, and Policy (www.science-society-policy.org).

Pelle Guldborg Hansen is Co-Director of ISSP – the Initiative for Science, Society and Policy. He has a Master's degree in Philosophy and Social Science from Roskilde University, Denmark, and obtained a PhD degree in Philosophy with a thesis on the game theoretical study of social conventions. Hansen has a broad interest in various issues at the intersection of social science, the Good life and technology; in particular, the study of the "choice architecture" and mechanisms of social systems and institutions.

Emily C. Parke is currently a PhD student in Philosophy at the University of Pennsylvania. She graduated from Reed College in 2000, and spent five years working for ProtoLife Inc. and co-organizing workshops at the European Center for Living Technology in Venice, Italy. She has co-authored several papers, and co-edited a book (*The prospect of protocells: Social and ethical implications of creating life from scratch*; MIT Press 2009), on the ethical implications of living technology and bottom-up synthetic biology.

Steen Rasmussen (Ph.D Technical University of Denmark, 1985) focuses mainly on pioneering and implementing new approaches, methods, and applications for self-organizing processes in natural and human made systems. He is currently the Director for the Center for Fundamental Living Technology (FLinT), Research Director at the Department for Physics and Chemistry at University of Southern Denmark (SDU), External Research Professor at the Santa Fe Institute (SFI), USA, Principle Investigator (PI) of the European Union (EC) sponsored Matrix for Chemical IT (MATCHIT) project and Co-PI for the EC sponsored Electronic Chemical Cell (ECCell) and Coordination of Biological & Chemical IT Research Activities (COBRA) projects. He was also the PI for the startup of the Initiative for Society, and Policy (ISSP) in Denmark, the Team Leader for the Self-Organizing Systems team at Los Alamos and a Guest Professor at University of Copenhagen (2004-5). He was PI for the Los Alamos Protocell Assembly (LDRD-DR) project and the Astrobiology program (origins of life) at Los Alamos, developing experimental and computational protocells and Cell-Like Entities, with USAF as a co-sponsor. Further, he was the co-director on the European Union sponsored Programmable Artificial Cell Evolution (PACE) project, and he was one of the founders of the Artificial Life movement in the late 1980s. He was the Chair of the Science and Engineering Leadership Team (SELT) for 2001-2002 in the Earth and Environmental Science (EES) Division at LANL and is currently on the Science Board for the European Center for Living Technology (ECLT) in Venice, Italy, which he co-founded in 2004. He also currently heads the Science Board for ProjectZero in Sønderborg, Denmark. Professor Rasmussen has published more than 85 peer reviewed papers and many internal technical reports, given more than 175 invited presentations outside of home institutions, and co-organized eight international and several national conferences. He organized the first two international protocell meetings, one at Los Alamos and the Santa Fe Institute (US) and one in Dortmund (Germany), and edited the first book on the topic. Many communications about his work inside and outside of the scientific establishment have appeared on television and in newspapers, periodicals, and books. Since 2000 he has sponsored 15 postdocs (theorists and experimentalists) and 30 graduate and undergraduate students. He is also actively engaged in the public debate about science, society and policy.